WIDE-FIELD TIME-DOMAIN FLUORESCENCE LIFETIME IMAGING MICROSCOPY (FLIM): MOLECULAR SNAPSHOTS OF METABOLIC FUNCTION IN BIOLOGICAL SYSTEMS

by

Dhruv Sud

A dissertation submitted in partial fulfillment
of the requirements for the degree of
Doctor of Philosophy
(Biomedical Engineering)
in The University of Michigan
2008

Doctoral Committee:

    Associate Professor Mary-Ann Mycek, Chair
    Professor David G. Beer
    Professor Nicholas Kotov
    Associate Professor Shuichi Takayama

UMI Number: 3343223

INFORMATION TO USERS

The quality of this reproduction is dependent upon the quality of the copy submitted. Broken or indistinct print, colored or poor quality illustrations and photographs, print bleed-through, substandard margins, and improper alignment can adversely affect reproduction.

In the unlikely event that the author did not send a complete manuscript and there are missing pages, these will be noted. Also, if unauthorized copyright material had to be removed, a note will indicate the deletion.

UMI Microform 3343223
Copyright 2009 by ProQuest LLC.
All rights reserved. This microform edition is protected against unauthorized copying under Title 17, United States Code.

ProQuest LLC
789 E. Eisenhower Parkway
PO Box 1346
Ann Arbor, MI 48106-1346

ACKNOWLEDGEMENTS

There are several people who's support made this thesis possible, and I would like to take a moment to acknowledge them.

First and foremost, I am thankful to my parents and siblings for their unwavering support during the course of my studies. My parents instilled in me the value of a good education very early on, something which has held me in good stead.

I would like to extend my gratitude to my Research Advisor, Prof. Mary-Ann Mycek, for her guidance. Her patience and encouragement while I was learning about the wide world of biomedical optics was key to the success of this work. Above all I appreciate her constant, constructive input; I am a better Researcher, Writer and Presenter for it.

I thank Professors David Beer, Nick Kotov and Shuichi Takayama for taking the time and effort to serve on my Dissertation Committee.

I am grateful to all the members of the Biomedical Optics Laboratory over the years, who made for a stimulating working environment. In particular, I would like to thank Dr. Wei Zhong, a Colleague and now a Research Fellow at Massachusetts General Hospital (Harvard Medical School), for aiding in all aspects of my research. Her attention to detail set a great example of research quality and thoroughness, which I follow to this day.

I would like to thank several UM colleagues for their collaborative work. Geeta Mehta/Prof. Shu Takayama for our work on imaging in bioreactors, and Khamir

Mehta/Prof. Jennifer Linderman for providing the Computational bioreactor model; Prof. David Beer for providing the human esophageal cell lines, and his constant input on the oxygen sensing and lysate studies; Elly Liao, Claire Jeong and Prof. Scott Hollister for our initial imaging studies on tissue engineered articular cartilage.

I would like to thank Dr. Periannan Kuppusamy (Ohio State University) for providing the LiNc-BuO particulates for EPR studies. I appreciate Arjun Khullar's early efforts on this project.

I also spend significant time under the guidance of Dr. Denise Kirschner before I started working with Mary-Ann. I thank her for introducing me to the mystery of Computational Pathology. The Kirschner Lab members remain friends till this day.

Joe Delli (Nuhsbaum) and Jonathan Girroir (Media Cybernetics) provided the Autoquant software for image restoration experiments, for which I am grateful as well.

I gratefully acknowledge financial support from the Whitaker Foundation and the National Institute of Health.

Last, but not the least, thanks to all my friends and colleagues for providing the social and recreational outlets that made my academic experience all the more wholesome.

TABLE OF CONTENTS

ACKNOWLEDGEMENTS ii

LIST OF FIGURES viii

LIST OF TABLES xii

ABSTRACT xiii

Chapter 1 INTRODUCTION 1

   1.1 Background and Motivation 1

      1.1.1 Biomedical Imaging 1

      1.1.2 Fundamentals of Fluorescence Microscopy 3

         1.1.2.1 Principles of Fluorescence 3

         1.1.2.2 Basics of Fluorescence Microscopes 6

      1.1.3 Fluorescence Lifetime Imaging Microscopy (FLIM) 8

         1.1.3.1 Overview of FLIM 8

         1.1.3.2 FLIM of Endogenous Fluorescence 11

         1.1.3.3 FLIM and Exogenous Fluorescence: Potential for Oxygen Sensing 13

         1.1.3.4 Noise Removal and Resolution Enhancement for FLIM 16

   1.2 Goals of this Work 20

   1.3 Dissertation Overview 21

Chapter 2 INSTRUMENTATION AND ANALYSIS 23

   2.1 Introduction 23

   2.2 FLIM 24

      2.2.1 Concept 24

2.2.2 Instrumentation 25

2.2.3 Rapid Lifetime Analysis 29

2.2.4 Noise Removal 31

Chapter 3 INTRACELLULAR OXYGEN SENSING IN LIVING CELLS 37

3.1 Introduction 37

3.2 Materials, Instrumentation, and Methods 40

3.2.1 Fluorescence Lifetime Imaging Microscope (FLIM) 40

3.2.2 Image Analysis 41

3.2.3 Temperature Control 41

3.2.4 Confocal Microscopy 42

3.2.5 RTDP Characterization 42

3.2.6 Calibration of Oxygen Sensitivity of RTDP 43

3.2.7 Cell Preparation 45

3.2.8 RTDP Lifetime Measurement in Cells 46

3.3 Results 48

3.3.1 NADH Measurements in HET and SEG 48

3.3.2 RTDP Calibration 50

3.3.3 Oxygen Measurements in HET and SEG 53

3.4 Discussion 57

Chapter 4 CALIBRATION AND VALIDATION OF INTRACELLULAR OXYGEN MEASUREMENTS 64

4.1 Introduction 64

4.2 Materials and Methods 67

4.2.1 FLIM and Oxygen Sensing 67

4.2.2. Cellular Lysate Generation and FLIM Analysis 67

4.2.3 EPR Oximetry 68

4.3 Results 70

4.3.1 RTDP-FLIM Analysis of Cellular Lysate 70

4.3.2 EPR Oximetry 71

4.4 Conclusion 72

Chapter 5   OXYGEN MONITORING FOR CONTINUOUS CELL CULTURE 75

5.1 Introduction 75

5.2 Materials and Methods 77

5.2.1 FLIM and Quantitative Oxygen Sensing 77

5.2.2. Bioreactor Fabrication and Cell Seeding 78

5.2.3. Bioreactor Imaging and Computational Validation 79

5.3 Results 79

5.3.1 Effect of Cell Density 80

5.3.2 Heterogeneity of Oxygen Distribution 81

5.4 Conclusion 83

Chapter 6 IMAGE RESTORATION IN FLIM 86

6.1 Introduction 86

6.2 Materials and Methods 90

6.2.1 Sample Preparation 90

6.2.2 Image Acquisition and Analysis 91

6.2.3 Image Restoration 91

6.3 Results 93

6.3.1 Computational Restoration of Gated Images 93

6.3.2 Weighted Intensity-Lifetime Mapping 95

6.4 Conclusion 98

Chapter 7 CONCLUSIONS AND FUTURE WORK 101

7.1 Conclusions 101

7.2 Future Experiments 107

7.2.1 Validation of Mitochondrial Dysfunction 107

7.2.2 Microfluidic Bioreactor Study 108

7.2.3 Lysis Study 108

7.2.4 FLIM Image Restoration 108

7.3 Future Applications 109

7.4 Closing Remarks 110

REFERENCES 113

LIST OF FIGURES

Figure 1. A typical Jablonski diagram (courtesy of Karthik Vishwanath). The singlet ground, first, and second electronic states are depicted by S_0, S_1, and S_2, respectively. Following light absorption, a fluorophore is usually excited to some higher vibrational level of either S_1 or S_2. With few exceptions, it rapidly relaxes to the lowest vibrational level of S_1 through internal conversion before it returns to the ground state S_0 via radiative or non-radiative decays.................4

Figure 2: Illustration of the microscope as a convolution operator: Convolution of the 'true' object image with the PSF yields the final image as seen by the observer. Image coutesy of Media Cybernetics....................7

Figure 3. Schematics of epi-illumination. Excitation light passes through the excitation filter and is reflected by a dichroic mirror. It then travels through the microscope objective to excite the sample. Fluorescence emission is collected with the same objective and passes through the dichroic mirror and the emission filter before it reaches the detector...................8

Figure 4. A cuvette of fluorescent dye excited by single photon excitation (top line, indicated by green arrow) and multiphoton excitation (localized spot of fluorescence, indicated by blue arrow) illustrating that two photon excitation is confined to the focus of the excitation beam (courtesy of Brad Amos MRC, Cambridge)...................17

Figure 5. FLIM concept. The system captures fluorescence intensity image at a time t_G after the excitation pulse over the interval Δt. Lifetime image can be created using intensity images captured at several different t_G [63]...................24

Figure 6: Fluorescence Lifetime Imaging Microscopy (FLIM) setup. Abbreviations: CCD—charge coupled device; HRI—high rate imager; INT—intensifier; TTL I/O—TTL input/output card; OD—optical discriminator. Abbreviations for optical components: BS—beam splitter; DC—dichroic mirror; FM—mirror on retractable 'flip' mount; L1, L2, L3, L4, L5—quartz lenses; M—mirror. Thick solid lines—light path; thin solid lines —electronic path. 25

Figure 7: Illustration of lifetime determination for a single pixel with the 2-gate RLD approach. 29

Figure 8: Illustration of adaptive and baseline noise removal for a fluorescent structure (inner green circle) with haze (outer green circle). G1,G2,G3,G4 = Gated images 1-4. The inset square area denote the user-defined haze region. Note how it changes intensity for each gated image for adaptive technique, but retains the same value for the baseline approach. 31

Figure 9: Comparison of adaptive (A) and baseline (B) approaches applied to images of RTDP-labeled living adenocarcinoma (SEG) cells. Line profiles are shown are shown for three lines as plots of lifetime (hundreds of ns, y-axis) and pixel number (x-axis), for both adaptive (green lines) and baseline (red lines) approaches. 34

Figure 10: A) Plot of RTDP intensity (pink squares) and lifetime (blue diamonds) vs. RTDP concentration. Lifetime was contant over a large range, while intensity varied with concentration before saturating. B) RTDP Calibration Curve: Plot of relative lifetime (τ_0/τ) vs. oxygen levels (μM). RTDP calibration indicated a linear relationship between oxygen levels and relative lifetime, which was in good agreement with the Stern-Volmer equation. Over multiple runs $K_q = 4.5 \pm 0.4 \times 10^{-3}$ μM^{-1}. The intercept $\neq 1$, indicating some degree of experimental variance. The calibration could differentiate between oxygen levels differing by as little as 8μM. 40

Figure 11: A) Oxidative phosphorylation in mitochondria. Complex I (NADH dehydrogenase) converts NADH to NAD^+ and passes on electrons to carrier CoenzymeQ (CoQ) while pumping hydrogen ions into the intermembrane space. Complex II (succinate dehydrogenase) can also generate and pass on electrons to CoQ via a complex internal mechanism which is initiated by conversion of succinate to fumarate, a step in the Krebs cycle. CoQ transfers the electrons to an intermediary complex III (CoQ – cytochrome C oxidoreductase), which in turns enhances the proton gradient by pumping hydrogen ions against the gradient. The electrons are transferred to complex IV (cytochrome oxidase) via cytochrome C (CytC) which in turn hydrolyses oxygen to water and also pumps protons against the gradient. The ATP synthase complex (complex V) moves protons down the gradient, converting osmotic energy to chemical energy via ATP synthesis from ADP in a process known as chemiosmotic coupling. A more detailed explanation of the process and the unique structure of ATP synthase can be found in any standard biochemistry text. B) NADH fluorescence and C) Mitotracker-stained SEG images. For the Mitotracker, excitation = 543 nm and emission = 636 nm. Mitotracker Red is a commercially available stain used for tracking mitochondria within living cells. Fluorescent signals from both markers were found to co-localize. Confocal intensity images of NADH fluorescence in HET (D) and SEG (E): Illustration of differences observed with the Zeiss 5 LSM. The SEG consistently presented a brighter signature than the HET by approximately 2.5-fold. F) Plot of differences in NADH fluorescence intensity and lifetime observed between the HET and SEG over multiple measurements with the FLIM system. No significant differences in NADH lifetime were observed. 49

Figure 12: (A) confocal fluorescence and (B) Differential Interference Contrast, or DIC images of SEG. SEG were incubated with the RTDP dye prior to imaging (see Methods section). FLIM images of SEG incubated with RTDP: (C) DIC, (D) fluorescence intensity in counts, (E) lifetime in ns and (F) oxygen in μM. Note that one cell in the bottom of (C) shifted position in (D) in the time lapse between these two images. (G) The results of depletion experiments on SEG (see Methods section). Cellular viability is compromised with the passage of time and this results in the lifetime / oxygen content leveling off towards the end. 54

Figure 13: RTDP fluorescence intensity (A,B), lifetime in ns (C,D) and oxygen in μM (E,F) maps of HET (A,C,E) and SEG (B,D,F). The intensity images (A,B) could not be reliably used to discriminate between the two different cell lines. The binary lifetime maps (C,D), on the other hand, plainly indicate different lifetimes for these two cellular species, with the SEG recording lower lifetimes than the HET. For the given case, $\tau_{HET} = 225$ ns and $\tau_{SEG} = 170$ ns. The mean lifetime difference was found to be $\Delta_\tau = 44 \pm 7.48$ ns. Logically, this translated into higher oxygen levels in the SEG vs. the HET using the calibration derived earlier, as can be seen in the oxygen distribution maps (C,F). G) illustrates the differences in oxygen levels within the HET and SEG as measured over multiple runs and assessed using the RTDP calibration. The mean difference between the two cell lines was hence evaluated as $\Delta_{[O]2} = 78 \pm 13$ µM and this value was statistically significant (p< 0.001)..57

Figure 14: Illustration of FLIM (imaging) and EPR (Spectroscopy) methods for quantitative oxygen sensing in living cells. For FLIM, RTDP lifetime images serve as the raw data. Oxygen sensitivity of RTDP can be calibrated via the Stern-Volmer equation and applied to every pixel in the lifetime image to yield an oxygen distribution image. 66

Figure 15. Illustration of EPR theory and operation. A) Energy absorption by the electron to shift between parallel (-1/2) and anti-parallel (+1/2) states results in a peak in the absorption spectrum, denoted as a plot of absorption (ab) vs. Magnetic Field (M). B) The derivation of the absorption spectrum is a measure of the slope and indicative of the environment of the spin probe. 69

Figure 16. Illustration of fluorescence intensity and lifetime imaging in microfluidic devices. Top: perspective view of a device that contained C2C12 mouse myoblasts and was perfused with media containing RTDP at a rate of approximately 0.5 nl/s by gravity-driven flows. Channel height = 50 µm, width = 300 µm. Bottom: representative images of RTDP fluorescence intensity (scale in counts), lifetime (microseconds), and oxygen (µM) obtained via FLIM..........................80

Figure 17. Simulation (white squares) and experimental FLIM (red squares) results of oxygen levels vs. cell densities in channels illustrated in Fig. 16. Oxygen levels were estimated by averaging pixel values in oxygen distribution images of the channel. The model simulations were carried out according to the equations described in [106], with the model parameters set at: maximum oxygen uptake rate $V_{max} = 2e^{-16}$ mol/cell/s; oxygen level at half-saturation $K_m = 0.0059$ mol/m^3; overall mass transfer coefficient $k_{la} = 4.5e^{-7}$ m/s, and estimated velocity of gravity flow <u> = 0.003 m/s. Error bars for some experimental data were within the red squares...................81

Figure 18. FLIM-based oxygen measurements from a closed-loop PDMS bioreactor for continuous cell culture of C2C12 mouse myoblasts. a) Device schematic. Channel shape was an isosceles trapezoid with a height of 30 µm and an upper (lower) PDMS layer of 180 µm (402 µm). Each of the six loops has a right and left valve separating it from the others. b) Oxygen distribution images at different points of a single loop (binary scale in µM). 82

Figure 19: Illustration of lateral smearing, or haze, with an image-intensified CCD camera. A) Blue fluorescence from a fixed mouse intestine section imaged with a CCD alone. B) Same region when imaged with an ICCD. The demagnification due to the lens-coupling between the intensifier and CCD (=2.17) is evident in the image. C) Red rectangular region from (B) magnified to show loss of resolution. The excitation source for all images was a mercury lamp. 88

Figure 20: Illustration of ICCD operation. The CCD camera is denoted by the CCD chip at the end of the diagram which provides the digital readout to the PC; all other components are part of the image intensifier. 89

Figure 21. A,B: Native fluorescence intensity, lifetime images of 3-micron diameter YG spheres. C,D: Corresponding restored images. The previously indistinguishable pair of spheres are evident in the restored lifetime image, as is a reduction in edge pixels with large lifetime values. 94

Figure 22. A,B: Native fluorescence intensity, lifetime images of RTDP-incubated SEGs. C,D: Corresponding restored images. 95

Figure 23. A: Native fluorescence intensity of five 10-micron YG beads. B. Native fluorescence lifetime map C: Restored intensity image. Note the reduction in haze. D. Restored intensity weighted-lifetime map..97

Figure 24. A: Low resolution (10x) native intensity image of a fixed mouse intestine section exhibiting Alexa 350 fluorescence. B. Native fluorescence lifetime map. C. Restored intensity image. D. Restored intensity weighted-lifetime map. 98

LIST OF TABLES

Table 1. Lifetime of commonly studied endogenous fluorophores [22]. Excitation (emission) denotes the wavelength value at the maximum of the excitation (emission) spectrum..................13

Table 2. Lifetime of commonly studied exogenous fluorophores [22]. Excitation (emission) denotes the wavelength value at the maximum of the excitation (emission) spectrum..................14

Table 3. Specifications of FLIM components. 28

Table 4: Lifetime differences and oxygen data from FLIM experiments on both living cells as well as for cellular lysates (data in grey boxes). All lifetime differences were computed as $\Delta\tau = \tau_{HET}-\tau_{SEG}$. Revised values of $\Delta\tau$ were computed as (a-b) and used to correct $[O_2]_{SEG}$ levels. K values were estimated via the Stern Volmer equation with known lifetime and corrected oxygen values for both HET and SEG cell lines...71

Table 5. List parameters used as input for computational image restoration. 92

ABSTRACT

Steady-state fluorescence imaging is routinely employed to obtain physiological information but is susceptible to artifacts such as absorption and photobleaching. FLIM provides an additional source of contrast oblivious to these but is affected by factors such as pH, gases, and temperature. Here we focused on developing a resolution-enhanced FLIM system for quantitative oxygen sensing. Oxygen is one of the most critical components of metabolic machinery and affects growth, differentiation, and death. FLIM-based oxygen sensing provides a valuable tool for biologists without the need of alternate technologies. We also developed novel computational approaches to improve spatial resolution of FLIM images, extending its potential for thick tissue studies.

We designed a wide-field time-domain UV-vis-NIR FLIM system with high temporal resolution (50 ps), large temporal dynamic range (750 ps – 1 μs), short data acquisition/processing times (15 s) and noise-removal capability. Lifetime calibration of an oxygen-sensitive, ruthenium dye (RTDP) enabled *in vivo* oxygen level measurements (resolution = 8 μM, range = 1 – 300 μM). Combining oxygen sensing with endogenous imaging allowed for the study of two key molecules (NADH and oxygen) consumed at the termini of the oxidative phosphorylation pathway in Barrett's adenocarcinoma columnar (SEG-1) cells and Esophageal normal squamous cells (HET-1). Starkly higher intracellular oxygen and NADH levels in living SEG-1 vs. HET-1 cells were detected by FLIM and attributed to altered metabolic pathways in malignant cells.

We performed FLIM studies in microfluidic bioreactors seeded with mouse myoblasts. For these systems, oxygen concentrations play an important role in cell behavior and gene expression. Oxygen levels decreased with increasing cell densities and were consistent with simulated model outcomes. In single bioreactor loops, FLIM detected spatial heterogeneity in oxygen levels as high as 20%.

We validated our calibration with EPR spectroscopy, the gold standard for intracellular oxygen measurements. Differences between FLIM and EPR results were explained by cell lysate-FLIM studies. We proposed a new protocol for estimating oxygen levels by using a reference cell line and cellular lysate analysis. Lastly, we proposed and compared two different image restoration approaches, direct lifetime vs. intensity-overlay. Both approaches improve resolution while maintaining veracity of lifetime.

Chapter 1 INTRODUCTION

1.1 Background and Motivation

1.1.1 Biomedical Imaging

Biomedical imaging refers to the techniques and processes used to create images of parts of the human body for clinical purposes or medical sciences. Imaging techniques form a critical tool for diagnosis and examination of diseases, as well as studying normal anatomy and physiology, in the modern healthcare scenario [1-5]. Integral to biomedical imaging is the non-invasive generation of images of internal aspects of the body. Prevalent approaches span the gamut of 2D and 3D imaging and include Computed Tomography (CT), Ultrasound, Magnetic Resonance Imaging (MRI), X-ray, Nuclear Medicine, Positron Emission Tomography (PET), Endoscopy and Microscopy [6-9]. While important and (often) complementary information is revealed by these techniques, most of them yield structural and anatomical information, and mostly at the tissue/organ scale (low resolution) [10]. In order to develop novel imaging-based techniques to aid in preventive and early stage diagnosis of life-threatening diseases such as cancer, it becomes imperative to understand these disorders at a molecular/cellular level, rather than the physical manifestation of the disease alone.

Optical imaging, or the response of biological systems to light as a means of contrast, is a promising technique. Contrast is based on spatial variation of properties such as absorption, scattering, reflection and fluorescence [11]. Despite the relative opacity of skin, light in the near-IR region has been demonstrated to penetrate deep into tissues [12, 13]. Typical penetration depths for commonly used lasers, such as Argon and Nd:YAG, are 0.5-2mm and 2-6mm, respectively [14]. Lastly, several organs are accessible via non-invasive or minimally invasive endoscopic systems for high resolution *in vivo* imaging [15-18].

In particular, fluorescence-based imaging methods provide a means of contrast not readily available with other optical techniques. Fluorescence imaging can provide both structural and chemical information down to the nanometer scale, thereby enabling single cell and even single molecule studies [19-21]. Hence, complex biological entities can be probed with high sensitivity and selectivity. Several parameters can be simultaneously probed such as spectra, quantum efficiency, intensity, lifetime and polarization [22].

From a biologic perspective, a variety of specimens exhibit self or autofluorescence (without the application of fluorophores) when they are irradiated, a phenomenon that has been exploited in the fields of botany, petrology, and even semiconductor biology [23, 24]. In contrast, the study of animal tissues and pathogens is often complicated with either extremely faint or bright, nonspecific autofluorescence. Of far greater value for the latter studies are exogenous fluorophores, which are excited by specific wavelengths and emit light of defined and useful intensity. Exogenous fluorophores are stains that attach themselves to visible or sub-visible structures, are often highly specific in their attachment targeting, and have a significant quantum yield (the ratio of photon absorption to emission). The widespread growth in the utilization of fluorescence microscopy is closely linked to the development of new synthetic and naturally occurring fluorophores with known intensity profiles of excitation and emission, along with well-understood biological targets.

1.1.2 Fundamentals of Fluorescence Microscopy

The concept of fluorescence was put forth by British scientist Sir George G. Stokes in 1852, and he was responsible for coining the term when he observed that the mineral fluorspar emitted red light upon illumination with ultraviolet excitation [25]. Stokes noted that fluorescence emission always occurred at a longer wavelength than the excitation light, a phenomenon labeled the Stokes Shift. Early investigations in the 19th century showed that many specimens (including minerals, crystals, resins, crude drugs, butter, chlorophyll, vitamins, and inorganic compounds) fluoresce when irradiated with ultraviolet light [14]. However, it was not until the 1930s that the use of fluorochromes was initiated in biological investigations to stain tissue components, bacteria, and other pathogens [14]. Several of these stains were highly specific and stimulated the development of the fluorescence microscope.

1.1.2.1 Principles of Fluorescence

Fluorescence is the result of a multi-stage process that occurs in certain compounds (prominently in polyaromatic hydrocarbons or heterocycles) called fluorophores or fluorochromes, as illustrated by a typical Jablonski diagram [22] (Fig. 1).

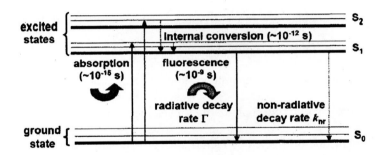

Figure 1. A typical Jablonski diagram (courtesy of Karthik Vishwanath). The singlet ground, first, and second electronic states are depicted by S_0, S_1, and S_2, respectively. Following light absorption, a fluorophore is usually excited to some higher vibrational level of either S_1 or S_2. With few exceptions, it rapidly relaxes to the lowest vibrational level of S_1 through internal conversion before it returns to the ground state S_0 via radiative or non-radiative decays.

A fluorophore is excited to the higher vibrational level of S_1 or S_2 following light absorption. With a few exceptions, it rapidly relaxes to the lowest energy level (S_1) through internal conversion in 10^{-12} seconds or less. At this level, the fluorophore is also subject to a multitude of possible interactions with its molecular environment. After this, a photon is released when the excited electron returns to the original ground state S_0.

Due to energy dissipation in these processes, the energy of the emitted photon is lower, and hence of longer wavelength, than the excitation photon. The difference in energy and wavelength is called the Stokes Shift [25]. The Stokes Shift is fundamental to the sensitivity of fluorescence techniques because it allows emission photons to be detected against a high background of excitation photons.

Fluorescence lifetime and quantum yield are two significant quantities associated with a fluorophore. Quantum yield is the number of photons emitted relative to the number absorbed. Since fluorescence decays radiatively (i.e. with the emission of photons) as well as non-radiatively (no emission), the quantum yield is always less than one. Fluorescence lifetime is defined by the average time the molecule spends in the excited state prior to returning to the ground state [26]. Lifetime determination is indicative of the time spent by the fluorophore in the excited state, which in turn depends on the surrounding and is affected by factors such as pH, dissolved gases, viscosity, binding, etc. Lifetime determination is therefore key for biological studies and for studying the biological milieu of the fluorophore.

For the two-state model shown in Fig. 1, a system with N fluorophores in the excited state depopulates stochastically at a rate:

$$dN(t)/dt = -(\Gamma + K_{nr})N(t) \quad (1)$$

where Γ is the radiative decay rate and K_{nr} is the nonradiative decay rate. The resulting decay is exponential in time with a lifetime τ given by:

$$\tau = 1/(\Gamma + K_{nr}) \quad (2)$$

Therefore, τ reflects the average time a molecule spends in the excited state.

The radiative decay rate Γ is dependent on the electronic properties of an isolated fluorophore, whereas the nonradiative decay rate K_{nr} is dependent on interactions between the fluorophore and its local environment, including mechanisms like dynamic or collisional quenching, molecular associations and energy transfer [22].

Fluorescence quenching is a process that reduces the fluorescence quantum yield without changing the fluorescence emission spectrum; it can result from transient excited-state interactions (collisional quenching), which reduces the fluorescence lifetime, or from formation of non-fluorescent ground-state species (static quenching), which does not affect the lifetime [22]. Note that lifetime is intrinsically not sensitive to factors affecting steady state intensity measurements such as fluorophore concentration and photobleaching.

1.1.2.2 Basics of Fluorescence Microscopes

Image resolution and contrast in the microscope can only be fully understood by considering light as a train of waves. Light emitted by a particular point on a specimen is not actually focused to an infinitely small point in the conjugate image plane, but instead light waves converge and interfere near the focal plane to produce a diffraction pattern. The ensemble of individual diffraction patterns spatially oriented in two dimensions, often termed Airy patterns, is what constitutes the image observed when viewing specimens through the eyepieces of a microscope.

The impulse response of the microscope can be recorded as the image of an infinitely small point source. Practically, this can be realized by imaging a fluorescent microsphere the size of which is below the resolution of the system, or theoretically from the numerical aperture of the objective lens, the refractive index of the sample, the pixel spacing on the CCD, and wavelength of light. The 2D Airy pattern, when observed in 3D, gives rise to the impulse response, or the point spread function (PSF) of the microscope. Every microscope image can then be (ideally) modeled as a convolution between the actual object and the PSF (see Fig. 2). In reality the PSF can be distorted by aberrations such as chromatic, spherical, blur, coma, astigmatism, etc. The basis of most computational image restoration techniques is the recovery of the original object image using the final image and knowledge of the PSF.

Figure 2: Illustration of the microscope as a convolution operator: Convolution of the 'true' object image with the PSF yields the final image as seen by the observer. Image courtesy of Media Cybernetics.

The basic function of a fluorescence microscope is to excite the specimen with a desired and specific band of wavelengths, and then to separate the much weaker emitted fluorescence from the excitation. In a properly aligned and configured microscope, only the emission should reach the detector, so that the resulting fluorescent structures are superimposed with high contrast against a very dark (ideally black) background. The limits of detection are generally governed by the darkness of the background and by the ability to eliminate excitation light, since it is typically several hundred thousand to a million times brighter than the emitted fluorescence.

Epi-fluorescence illumination (Fig. 3) is the overwhelming choice of configuration in modern microscopy. The illumination is designed to direct light onto the specimen by first passing the excitation light through the microscope objective onto the specimen, and then using that same objective to capture the emitted fluorescence [14]. This type of illuminator has several advantages. The microscope objective serves first as a well-corrected condenser and secondly as the image-forming light gatherer. Being a single component, the objective/condenser is always in perfect alignment. A majority of the excitation light reaching the specimen passes through without interaction and travels away from the objective, and the illuminated area is restricted to that which is observed through the eyepieces (in most cases). Unlike the situation in some contrast enhancing techniques, the full numerical aperture of the objective is available when the microscope is properly configured. If desired, it is possible to combine with or alternate between reflected light fluorescence and transmitted light observation and the capture of digital images.

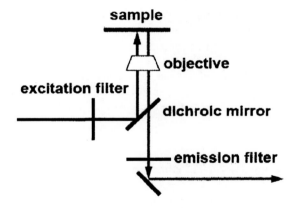

Figure 3. Schematics of epi-illumination. Excitation light passes through the excitation filter and is reflected by a dichroic mirror. It then travels through the microscope objective to excite the sample. Fluorescence emission is collected with the same objective and passes through the dichroic mirror and the emission filter before it reaches the detector.

1.1.3 Fluorescence Lifetime Imaging Microscopy (FLIM)

1.1.3.1 Overview of FLIM

While methods of steady-state fluorescence microscopy are widely used in biology and medicine to reveal information on anatomical features, cellular morphology, cellular function, and intracellular microenvironments, measurements of fluorophore excited-state lifetimes offer an additional source for contrast for imaging applications because fluorescence lifetimes are highly sensitive to physical conditions in the fluorophores local environment, such as temperature, pH, oxygen levels, polarity, binding to macromolecules and ion concentration, while being generally independent of factors affecting steady-state measurements such as fluorophore concentration, photobleaching, absorption and scattering.

Fluorescence lifetime imaging microscopy (FLIM), as a method for producing spatially resolved images of fluorescence lifetime, was first introduced in 1989 [27]. Fluorophore lifetimes can be measured in the time domain, where the system's impulse response to pulsed excitation is probed, or the frequency domain, in which the system's harmonic response to a modulated excitation is measured [28, 29].

Measuring lifetime (t) via TD is more intuitive. It exploits the fact that the fluorescence emission is theoretically proportional to the number of molecules in the first excited state, and hence it decays exponentially. The exponential decay can be reconstructed in different ways, most commonly used of which are gated integration and time-correlated single photon counting (TCSPC). For FD FLIM, a sinusoidally modulated light source is used for excitation. The resulting sample emission is also sinusoidally modulated at the same frequency as the excitation, but is shifted in phase and is demodulated to some extent, that is has a reduced modulation depth. Fluorescence lifetime can be directly calculated by changes in phase delay and demodulation. For measuring different lifetimes, different modulating frequencies need to be used. A more detailed description of FD FLIM can be found elsewhere [30].

In principle, time- and frequency-domain (TD and FD) methods are equivalent and related by a Fourier transformation [11]. Historically, FD FLIM had been easier to implement due to the ability to extract 100 MHz information by means of cross-correlated detection, eliminating the need for fast electronics. Low signal strengths and slow image processing times were major disadvantages, though currently, real-time *in vivo* FD FLIM systems have been developed [31, 32]. By comparison, TD FLIM had been harder to implement due to the lack of available femto- and picosecond pulsed laser sources and the difficulty in implementing subnanosecond gated detectors. The development of femtosecond Ti:sapphire sources and high-power (>100 mW), picosecond pulsed diode lasers, along with picosecond gated intensified CCD detectors, has eliminated major technological difficulties and is implemented in TD systems [31, 33-35]. Presently, both methods have comparable temporal resolution and discrimination, and both benefit from rapidly advancing technologies. Therefore, the choice of FLIM implementation depends on the specific application, availability of experimental equipment, and the nature of the lifetime information to be extracted. TD methods have a greater temporal dynamic range and are better suited for detection of long lifetimes [30].

FLIM has been successfully applied in biology and biomedicine for determination of spatial ion and metabolite distributions, monitoring of interactions between cellular components by fluorescence/Förster resonance energy transfer (FRET), and for detection of abnormal tissues [36-39]. The simplicity of the FLIM method makes it attractive for many applications that may otherwise require tedious calibration processes to minimize artifacts that influence steady-state intensity measurement.

For example, FLIM was reported for quantitative pH determination in living cells and contrasted with the traditional ratiometric technique. In the FLIM method, the different lifetimes of the protonated and ionized forms of the probe (BCECF-AM) revealed intracellular pH [40]. FLIM was also applied to phthalocyanine photosensitizer distribution measurement in V79-4 Chinese hamster lung fibroblast cells [41]. The detailed lifetime image obtained was used to distinguish between inhomogeneous distributions of photosensitizers and localized intracellular quenching. The distinction between the two processes could only be made with lifetime imaging and would be impossible with conventional fluorescence intensity-based microscopy.

1.1.3.2 FLIM of Endogenous Fluorescence

An important application of FLIM is the imaging of endogenous fluorescence in cells and tissues. Endogenous fluorophores are considered as potential probes of metabolic function, tissue morphology and, therefore, have potential diagnostic importance in medicine with tissue fluorescence lifetime being a potential source for contrast. The diagnostic potential of lifetime-based imaging has been repeatedly demonstrated by distinguishing lifetime differences between normal and diseased states [42-45]. Lifetime studies on the endogenous fluorescence of human skin revealed variations in metabolic states of the cells and indicated a possibility of using FLIM for dermatological diagnosis of basal cell carcinoma [46]. Because exogenous agents are not employed, endogenous fluorescence methods raise no concerns regarding issues of contrast agent toxicity or delivery, shortening FDA approval times as well. Fluorescence lifetime information complements minimally invasive steady-state endogenous fluorescence methods for disease detection and metabolic imaging.

Endogenous fluorophores found in cells and tissues include amino acids (tryptophan, tyrosine, phenylalanine), metabolic cofactors (oxidized flavins, reduced pyrindine dinucleotides), structural proteins (collagen, elastin, keratin), vitamins (retinols, pyridoxines, riboflavins), lipids (lipofuscin, ceroid), and tetrapyrroles (porphyrins, chlorophylls). Table 1 lists optical properties of commonly studied endogenous fluorophores. Note that NADH (nicotinamide adenine dinucleotide or reduced pyridine nucleotides) and NADPH (nicotinamide adenine dinucleotide phosphate) have almost identical excitation-emission spectra. The abbreviation NAD(P)H is often used to emphasize the spectral indistinguishability of NADH and NADPH in cellular systems.

Fluorophore	Excitation (nm)	Emission (nm)	Quantum Yield	Mean Lifetime(ns)
Tryptophan	295	353	0.13	3.1
Tyrosine	275	304	0.14	3.6
Phenylalanine	260	282	0.02	6.8
NAD(P)H	350	460	0.05	0.4
Flavins:	450	525		
FAD			0.03	4.7
FMN			0.25	2.3
Collagen, type 1	325	400	0.40	5.3
Elastin, bovine neck ligament	350	410	0.09	2.3

Table 1. Lifetime of commonly studied endogenous fluorophores [22]. Excitation (emission) denotes the wavelength value at the maximum of the excitation (emission) spectrum.

1.1.3.3 FLIM and Exogenous Fluorescence: Potential for Oxygen Sensing

While endogenous fluorophores have several benefits, they often suffer from poor quantum efficiency, overlapping excitation and emission spectra, and limited applicability. Since the discovery of fluorescence, persistent efforts have been made to create fluorescent dyes with specific properties of excitation/emission, high quantum yield, photostability, specificity/sensitivity to certain biological structures/molecules, and low photoxicity/cytotoxicity. Several are variants of naturally occurring fluorophores in the animal kingdom (e.g. Lucifer yellow and green fluorescent protein, or GFP), though most are engineered. Table 2 is an illustration of some commonly used exogenous fluorophores and their optical properties.

Fluorophore	Excitation (nm)	Emission (nm)	Quantum Yield	Mean Lifetime(ns)
Fluorescein	495	519	0.93	3.25
Lucifer yellow	425	528	0.21	7.81
DAPI	345	455	0.58	0.2-2.8
EGFP	489	508	0.60	2.8
EYFP	514	527	0.61	3.27
Hoechst 33342	343	483	0.66	0.35-2.21
RTDP	460	600	0.04	350
Rose Bengal	540	550-600	0.05	0.09
Rhodamine 6G	526	555	0.95	4.08

Table 2. Lifetime of commonly studied exogenous fluorophores [22]. Excitation (emission) denotes the wavelength value at the maximum of the excitation (emission) spectrum.

Combining studies of exogenous and endogenous fluorescence provide a unique perspective into biological function that is unavailable with either approach alone. While such studies are usually *ex vivo* in nature, they provide important insight that can later be translated into a more clinic-friendly approach. A significant area of interest for such an approach is metabolic function, especially in tumor cell models [47, 48]. Metabolism in cancer cells is invariably linked to oxygen consumption (or the lack of it), and the oxidative phosphorylation (OXPHOS) chain in the mitochondria [49, 50]. Fortunately, endogenous fluorophores, such as NAD(P)H and FAD, are important metabolic cofactors and part of OXPHOS. In fact some of the early work on endogenous NAD(P)H fluorescence was motivated by the discovery that NAD(P)H signatures of normal and diseased tissue (e.g. tumors) were different [51].

Oxygen in living systems governs growth, differentiation and death, and hence forms an important area of study. Older approaches for biological oxygen sensing include Clark-type electrochemical electrodes and colorimetry [52, 53]. More recent approaches are electron paramagnetic resonance (EPR) and nuclear magnetic resonance (NMR) [54-57]. Fluorescence-based methods for oxygen sensing gained notice in the mid-1990s and offer the opportunity for non-invasive measurements with the high sensitivity and spatial resolution required for intracellular oxygen sensing, which remains a challenge. Most other methods assess intracellular oxygen levels by measuring extracellular oxygen concentration, under the assumption that oxygen levels remain the same throughout the cells, which may not always be valid. Furthermore, the accuracy of oxygen consumption measurements, when applied to groups or suspensions of cells, is limited by factors such as cell counting accuracy. In addition, there are inherent uncertainties associated with some oxygen-sensing methods, such as Clark-type electrodes, which consume oxygen during the measurement [58]. The current gold standard for intracellular oxygen sensing remains EPR, a spectroscopic method that is not routinely accessible to biologists.

Intracellular fluorescence lifetime imaging of ruthenium-based dyes has been reported using FLIM [59, 60]. One major advantage of fluorescence lifetime methods is their insensitivity to fluorophore concentration, thereby minimizing artifacts if oxygen-sensitive probes are heterogeneously distributed within the biological sample. While ruthenium-based oxygen sensing shows promise, many current FLIM systems do not temporal dynamic ranges large enough to image long fluorescence lifetime dyes. Since most biological fluorophores have lifetimes in the $10^{-9} - 10^{-8}$ s range, as compared with 10^{-6} s for ruthenium-based dyes, FD FLIM systems often use an excitation source modulated in the GHz range to optimize the detection sensitivity of the system to nanosecond lifetime fluorophores. However, FD FLIM systems may be limited in temporal dynamic range. TD FLIM systems have a greater dynamic range, though the increasingly common use of modelocked Ti:Sapphire lasers operating at 110 MHz makes long lifetime measurements difficult without the use of cavity dumpers or pulse pickers to lower the repetition rate. Therefore, high repetition rate systems may not be optimized for sensing both nanosecond lifetime fluorophores and microsecond lifetime ruthenium-based dyes. The use of tunable nitrogen lasers circumvents several of these issues. Nitrogen (UV) lasers for microscopy-based studies provide better correlation with clinical data since several clinical systems that employ laser-induced fluorescence (LIF) [61-63]. The lower repetition rate and single-shot ability of N_2 lasers also allows for imaging of long-lived fluorescence and even phosphorescence. Several ruthenium and platinum-based indicators exist with lifetimes on the order of microseconds and find wide application for oxygen sensing; nitrogen laser based FLIM is readily applicable for such purposes [64-66].

1.1.3.4 Noise Removal and Resolution Enhancement for FLIM

In theory, the spatial resolution of a conventional microscope is limited by diffraction to about 0.2 µm in the lateral direction and about 0.6 µm in the axial direction [67, 68]. In practice, the theoretical resolution of optical microscopy, however, is never achieved, because scattering by thick and turbid biological media causes light to enter the focal plane from above and below, thus blurring images [69]. Noise and further distortion in FLIM instrumentation itself can arise from both optical and digital sources. For example, nitrogen gas lasers exhibit intensity and temporal jitter during laser discharge, and ICCDs commonly have thermal noise, read-out noise, flicker noise, quantum noise and image degradation [70].

The past two decades have seen spectacular advances in fluorescence microscopy, with fundamental innovations in instrumentation for obtaining high resolution images, which provide new insights into organization and function of biological systems. The most prominent of these are the practical implementation of various forms of confocal microscopy and the invention of multiphoton microscopy [68]. In confocal microscopy, the essential component is a pinhole, which is conjugated to the focal point in the sample.

As a laser beam rapidly scans across a sample, the resulting fluorescence travels through the pinhole, which rejects defocused light before it reaches the detector. This means that the only light detected comes from a thin section of the object near the focal plane. This process is known as optical sectioning.

Figure 4. A cuvette of fluorescent dye excited by single photon excitation (top line, indicated by green arrow) and multiphoton excitation (localized spot of fluorescence, indicated by blue arrow) illustrating that two photon excitation is confined to the focus of the excitation beam. Image courtesy of Brad Amos MRC, Cambridge.

Multi-photon microscopy is another optical sectioning technique that uses infrared (IR) light to excite fluorophores usually excited by ultra-violet (UV) or visible light [68]. Multi-photon excitation works by using femtosecond pulses of low energy light to excite fluorophores. As the laser beam is focused, the spatial density of photons increases, and the probability of two of them interacting simultaneously to excite a single fluorophore increases. The laser focal point is the only location along the optical path where the photons are crowded enough to generate significant occurrence of two-photon excitation. The fall-off of photon density outside the focal volume is so steep that nothing outside of it is excited. Thus, optical sectioning with multi-photon imaging is intrinsic to the excitation process without any need for confocal apertures (see Fig. 4).

Both confocal microscopy and multiphoton microscopy produce images with finer details than conventional wide-field microscopy. Resolution in multiphoton microscopy does not exceed that achieved with confocal microscopy and, in fact, the utilization of longer wavelengths (red to near-infrared; 700 to 1200 nanometers) results in a larger point spread function for multiphoton excitation [71]. This translates into a slight reduction of both lateral and axial resolution. In practice, confocal resolution can be degraded by the finite pinhole aperture, chromatic aberration, and imperfect alignment of the optical system, all of which serve to reduce resolution differences between confocal and multiphoton microscopy. If structures are not adequately resolved with a confocal microscopy, they will not fare any better (and may be worse) when imaged with multiphoton excitation.

With confocal microscopy, although fluorescence is excited throughout the sample illuminated volume, only signal originating in the focal plane passes through the confocal pinhole, resulting in reduced signals. Moreover, the large excitation volume can cause significant photobleaching and phototoxicity problems, especially in live specimens. With multiphoton microscopy, the sample penetration depth is deeper than with both confocal microscopy and conventional wide-field microscope because IR light is less like to be absorbed and scattered by turbid tissue samples than UV light. Moreover, photobleaching and photodamage is minimized in multiphoton microscopy because there is no absorption in out-of-focus sample areas. However, because the photophysics governing two-photon excitation is different from that of conventional fluorescence excitation, deleterious effects are occasionally observed with two-photon excitation of certain fluorophores, which in turn limits the applicability of this method for optical sectioning in thin specimens. In addition, the instrumentation, particularly the lasers required for this technique are very specialized and expensive.

Although conventional wide-field microscopy does not offer as high resolution as confocal microscopy and multiphoton microscopy, it allows capturing images of the entire object field simultaneously in a parallel processing system, as compared to the scanning-based confocal microscopy and multiphoton microscopy. The whole-field nature of this approach can provide a very high data acquisition rate, making it attractive for imaging transient biological processes [72].

The classical challenge in wide-field microscopy is to reduce defocused light in order to obtain resolutions comparable to those achieved via confocal microscopy and multiphoton microscopy [73]. With FLIM systems that use image intensifiers, further resolution loss is observable due to smearing of the image (discussed in later chapters) [74, 75]. Image intensifiers are, however, a necessary evil for imaging of weakly fluorescent molecules, such as NADH and collagen.

Surprisingly little work has been done for improving spatial resolution of fluorescence lifetime images. Optical approaches such as structured illumination (SI) were recently demonstrated to improve resolution and reduce temporal blurring in FLIM [74]. SI and other optical approaches, however, result in a reduction of SNR in acquired images, which is detrimental for low-light imaging. Among computational approaches, a single paper describing image restoration in a frequency domain FLIM system was found [76]. Arguments against computational methods for FLIM analysis include lack of quantitative options and long processing times [77, 78]. With recent advances in development of constrained algorithms and improved desktop computing ability, it now becomes possible to consider such approaches for FLIM. Quantitative image restoration provides several benefits over optical approaches, such as improvement of SNR, no need for additional hardware, fidelity of lifetime information, widespread availability of research and commercial tools, and applicability to previously acquired data.

1.2 Goals of this Work

The specific aim of this work was **to develop fluorescence lifetime imaging microscopy (FLIM) with computational image restoration capability as an effective tool for accurate studies of metabolic function via combined endogenous (NADH) and exogenous (oxygen) sensing.**

To achieve this aim, the following steps were performed:
1. Development of a wide-field time-domain FLIM system with wide spectral tunability (UV-vis-NIR), large temporal dynamic range (750 ps – 1 μs), high sensitivity (50 ps), sample perfusion and temperature control, and with noise removal capability.

2. Characterization and development of a calibration procedure for the oxygen sensitivity of ruthenium tris(2,2'-dipyridyl) dichloride hexahydrate (RTDP) fluorescence lifetime, and to apply it to living cells.

3. Exhibit the potential for oxygen sensing for high resolution studies (single cell) of metabolic function in normal and diseased cells.

4. Validate oxygen measurements with electron paramagnetic resonance (EPR); correct by accounting for intracellular biochemical information via lysate studies. Propose generic protocol for accurate intracellular oxygen measurements, extensible to any probe.

5. Demonstrate wide applicability of quantitative oxygen sensing by applying it to oxygenation imaging in microfluidic bioreactors containing living cells in a 3D matrix.

6. Develop FLIM spatial resolution enhancement techniques via the 2D blind image restoration technique. Demonstrate effectiveness on various realistic biological samples.

1.3 Dissertation Overview

Chapter 2 describes wide-field time-domain FLIM instrumentation, data analysis, and various noise removal techniques.

Chapter 3 described application of RTDP-FLIM for intracellular oxygen sensing, and to living cancer cell models in particular.

Chapter 4 details the validation and calibration of the RTDP-FLIM oxygen sensing approach, including results from lysate and EPR studies.

Chapter 5 describes application of the RTDP-FLIM approach to extracellular oxygen sensing in microfluidic bioreactors.

Chapter 6 details our work on reducing noise and on resolution enhancement in FLIM via computational means.

Chapter 7 concludes this dissertation, describes potential future experiments and mentions possible applications.

Chapter 2 INSTRUMENTATION AND ANALYSIS

2.1 Introduction

Developments in imaging technologies allow researchers to extract information from biological systems, including cell morphology, intracellular ionic concentrations, and membrane integrity [79]. In particular, fluorescence lifetime studies have gained prominence as molecular timers that can be used to study a plethora of cellular events. Lifetime is an inherent property of the excited electronic state of the fluorescent molecule (fluorophore) and is defined as the average time the fluorophore spends in the excited state before returning to the ground state. While lifetime is independent of fluorophore concentration, photobleaching, absorption, and scattering, it is influenced by the local microenvironment of the fluorophore (e.g., pH, ions such as Ca^{2+}, molecular associations) and hence can be used to probe the intracellular milieu [80]. While fluorescence lifetime imaging has been put to a variety of uses, the most common has been for studying energy transfer during molecular associations in living cells as a function of intermolecular distance, thereby measuring molecular gaps below the resolution of current imaging capabilities [81, 82].

In this chapter we describe a wide-field, time-domain fluorescence lifetime imaging microcopy (FLIM) system that was developed to probe cellular metabolic function and detect molecular activity in living cells [70]. We introduce lifetime image analysis via the Rapid Lifetime Determination (RLD) method. Finally, we present two different approaches to noise removal, baseline and adaptive.

2.2 FLIM

2.2.1 Concept

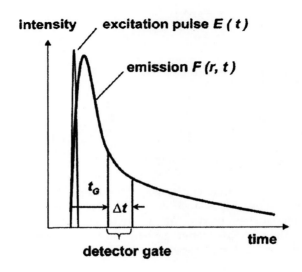

Figure 5. FLIM concept. The system captures fluorescence intensity image at a time t_G after the excitation pulse over the interval Δt. Lifetime image can be created using intensity images captured at several different t_G [70].

The concept of gated FLIM imaging is shown in Fig. 5. An excitation pulse E(t) illuminates the sample. The fluorescence emission is collected by the microscope and imaged by an intensified CCD (ICCD) camera at a controllable delay t_G with a gate width (similar to 'shutter time' for cameras) of Δt. Fluorescence lifetime is then determined by obtaining intensity images at several different t_G (gate delays) and fitting the intensity values to an exponential decay. When done for each pixel in an image, this results in a

FLIM image, where contrast is based solely on fluorescence lifetime. Fitting method(s) are discussed later in the chapter.

2.2.2 Instrumentation

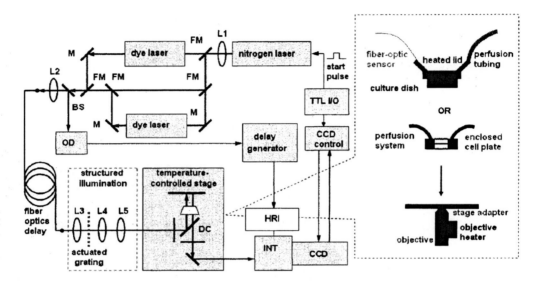

Figure 6: Fluorescence Lifetime Imaging Microscopy (FLIM) setup. Abbreviations: CCD—charge coupled device; HRI—high rate imager; INT—intensifier; TTL I/O—TTL input/output card; OD—optical discriminator. Abbreviations for optical components: BS—beam splitter; DC—dichroic mirror; FM—mirror on retractable 'flip' mount; L1, L2, L3, L4, L5—quartz lenses; M—mirror. Thick solid lines—light path; thin solid lines—electronic path.

Our wide-field time-domain FLIM system is illustrated in Fig. 6. Excitation light was coupled into a 20 m long optical fiber (SFS600/660N, Fiberguide, Stirling, NJ). The light exiting the fiber was projected into the back port of a Zeiss Axiovert S100 inverted microscope. A dichroic mirror delivered the beam into a 10x, 40x, or 100x Fluar (Zeiss, Jena, Germany) objective for sample excitation. Artifacts due to temporal jitter in the laser discharge were minimized by using a reference beam split from the main laser as the timing reference via an electronic pulse generated by means of a constant fraction optical discriminator (OD) (OCF-400, Becker&Hickl GmbH, Berlin, Germany), the output from which was sent to a picosecond delay generator (DEL350, Becker & Hickl GmbH) that provided an adjustable time delay to trigger an ultrafast gated (min. 200 ps, maximum jitter of 10 ps), intensified CCD camera (Picostar HR, La Vision, Goettingen, Germany) that was used for image detection. It is important to note the need of the optical fiber, that performs a vital optical delay function for FLIM. Since fluorescence lifetimes are typically a few nanoseconds, the ICCD shutter must open within picoseconds after sample excitation to collect the fluorescence decay. However, intrinsic delays arise due to electronic and cabling propagation of signal. The 20-m optical fiber hence serves as a time delay to bring the timing of fluorescence emission within an accessible measurement regime for the delay card and ICCD.

For temperature control, the Delta T controlled culture dish system (Bioptechs, Butler, PA) was implemented along with the Delta T perfused heated lid (Bioptechs) and was used for all experiments, unless noted otherwise. The heated lid had two ports which could be used for perfusion purposes. A remote controller was used to adjust temperature settings. For closed chamber experiments the FCS2 chamber system and controller (Bioptechs) was used, where cells were enclosed between a cell plate and a perfusable gasket with a media holding capacity of 1 ml. Due to the large thermal mass of the objective, a separate Objective Heater (Bioptechs) with its own controller was also installed. All temperature control systems had an operating range from ambient to 45°C with accuracy within ±0.2°C.

The FLIM system can excite in the UV-NIR range, from 337-960 nm depending on the laser dye used. The nitrogen laser is a pulsed source with peak energy of approximately 1.3 mJ with reproducibility within ± 2%. FLIM has a spatial resolution of 1.4 μm and (with structured illumination) can achieve an optical section down to 10 μm. A key advantage of this FLIM implementation is its ability to measure lifetimes of long-lived fluorophores, ranging from 750 ps to almost 1 μs with a resolution of 50 ps. Being a wide-field system where the entire image is captured simultaneously, our FLIM is capable of rapid image acquisition. This is advantageous for future clinical applications, where sample motion demands quick image capture capability. Finally, as indicated in Fig. 6, two different culture dish systems provide temperature controlled, perfused units for cellular study under more physiologically-apt conditions. Complete specifications of various FLIM components are listed in Table 3 [83].

The spatial resolution of FLIM images is dependent on the intensifier, which is listed as having a resolution of 35lp/mm. This resolution, however, only applies in the DC mode. In the gated mode, the resolution drops to 10-15lp/mm due to reduced voltage in the photocathode-MCP gap (lowered from 400V to 25V). This is mostly to prevent severe heating effects due to switching 400V at the maximum repetition rate (110 MHz) of the ICCD. The spatial resolution of fluorescence intensity images was determined to be approximately 1 μm [70]. More details on loss of spatial resolution in ICCDs and some possible solutions are presented in Chapter 6.

Component	Parameter	Value
Intensifier	Spectral Range (>4% QE)	300-650 nm
	Gate Widths	200 ps – 10^{-3} s
	Microchannel Plate Voltage	260-800 V
	Resolution	35 lp/mm
CCD	Spectral Range (>4% QE)	300-880 nm
	Number of Pixels	640 x 480
	Pixel Size	9.9 µm x 9.9 µm
	Dynamic Range	12 bit
	Readout Rate	12.5 MHz
	Readout Noise	2 counts @ 12.5 MHz
	Frame Rate	30 frames/s
	Exposure Times	1 ms – 1000 s
	Full Well Capacity	35,000 electrons
	Dark Current	<0.1 electrons/pixel/sec
	A/D Conversion Factor	7.5 electrons/count
Optical Discriminator	Spectral Range	300-1000 nm
	Time Walk (1 ns input)	<30 ps rms
Delay Generator	Delay Range	10 ns – 100 µs
	Absolute Delay Accuracy	5%
	Delay Resolution	12 bit, 2.5 ps resolution
	Delay Jitter	0.05% of range, 10 ps minimum
Optical Fiber	Material	Silica
	Core/Clad Diameter	600/660 µm
	Numerical Aperture	0.22
	Attenuation Factor	0.25 dB/m at 337 nm
	Refractive index (at 633 nm)	Core: 1.457, clad: 1.44

Table 3. Specifications of FLIM components [83].

2.2.3 Rapid Lifetime Analysis

FLIM imaging and analysis was done via an intensified CCD camera controlled by the DaVis 6 software (LaVision, Goettingen, Germany). A dark background was taken to account for thermal current in the CCD and was subtracted from each intensity image. The CCD camera is operated within its specified linear intensity-response range, which was controlled by two separate settings: 1) Adjusting the gain on the intensifier via the HRI and 2) modulating the gate width via the software interface. No adjustments were made to the intensity images other than those during image capture.

Since fluorescence lifetimes must be calculated on a per pixel basis for FLIM, iterative algorithms can be time-consuming and computationally intensive. Alternate approaches such as Rapid Lifetime Determination (RLD) are more amenable for imaging systems where speed/time is of value. The RLD approach in its simplest form (2-gates) is shown in Fig. 7.

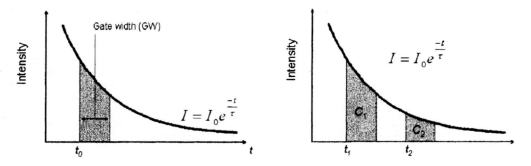

Figure 7: Illustration of lifetime determination for a single pixel with the 2-gate RLD approach.

Assuming an extremely short excitation pulse and a single exponential decay τ, imaged with an ICCD gate width GW, the detected photon counts at any pixel is given by:

$$counts(C) \approx \int_{t0}^{t0+GW} I_0 e^{-t/\tau} dt = \left[-I_0 \tau . e^{-t/\tau}\right]_{t0}^{t0+GW} = C_1 \qquad (3)$$

Two consecutive points, given by total counts C_1 and C_2 (Fig. 7), then define the single exponential decay:

$$\tau = \frac{t_2 - t_1}{\ln\left(C_1/C_2\right)} \qquad (4)$$

2-gate schemes are susceptible to noise, so we extend our acquisition to multiple gates and implement a version of the RLD that is extensible to an arbitrary number of gates. Lifetime maps were calculated from multiple intensity maps on a per-pixel basis by fitting the lifetime of pixel p to the logarithm of the intensity (assuming single-exponential decay):

$$\ln I_{i,p} = -\frac{t_i}{t_p} + C \qquad (5)$$

where $I_{i,p}$ is the intensity of pixel p in image i, t_i is the time delay of image i, and C is a constant. Least squares lifetime fits were:

$$\tau_p = -\frac{N\left(\sum t_i^2\right) - \left(\sum t_i\right)^2}{N\sum t_i \ln I_{i,p} - \left(\sum t_i\right)\left(\sum \ln I_{i,p}\right)} \qquad (6)$$

where N was the number of images. All sums are over i. At each delay, the average of five intensity images was recorded to compensate for excitation laser intensity. All images were pre-processed via a uniform threshold filter where pixels below a minimum, non-zero intensity value were set to zero. The nitrogen laser was operated at about 3 Hz and 640×480 pixel lifetime images using 4 gates were generated using a PC with an Intel Pentium-IV processor at 3.2 GHz, for total acquisition and analysis times <15 s. Accurate lifetimes were obtained by creating a histogram based on the lifetime map, fitting it to log-normal functions and calculating the average time from the regression. The resulting image is the final lifetime map.

2.2.4 Noise Removal

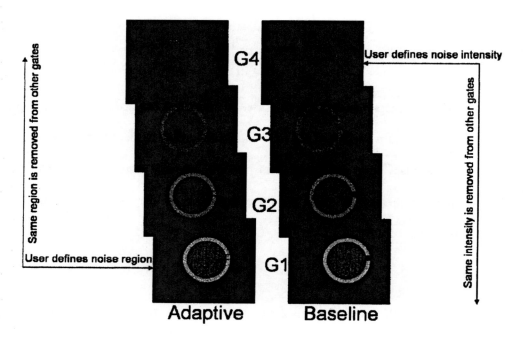

Figure 8: Illustration of adaptive and baseline noise removal for a fluorescent structure (inner green circle) with haze (outer green circle). G1,G2,G3,G4 = Gated images 1-4. The inset square area denote the user-defined haze region. Note how it changes intensity for each gated image for adaptive technique, but retains the same value for the baseline approach.

ICCDs form an important component of low-light imaging systems, but also lead to loss of spatial resolution. One issue with ICCD use is the lateral smearing of fluorescence, which manifests in the form of haze around fluorescent structures. We have already described how thermal and background noise is removed during image acquisition and processing; here we describe two additional approaches which allow the User to pick and remove haze as much as possible for 2D imaging. More sophisticated noise removal for

3D imaging via image restoration will be discussed in later chapters. Adaptive and baseline noise removal is illustrated in Fig. 8.

4 gated images (i.e. taken at different delays) were acquired via FLIM and analyzed to yield lifetime maps as described previously [70]. The lifetime macro was modified to implement both adaptive and baseline haze removal.

a) Adaptive – The user was prompted to select a rectangular region representative of haze in the first gated image (highest SNR). For *each* gated image then, the macro averaged the pixel intensity in the selected rectangular area and subtracted the average value on a per pixel basis from that gated image alone. The term 'adaptive' hence comes from the fact that the haze removal from each gated image is dependent on that particular image alone.

b) Baseline – The baseline approach is similar to the adaptive, except that the User selected a haze area from the last gated image (lowest SNR), which was then averaged and subtracted on a per pixel basis from ALL gated images. This approach is hence equivalent to a single offset value applied to all gated images.

Since haze is almost always greater than background noise, both adaptive and baseline methods also serve to remove background and, to some extent, dim defocused fluorescence as well.

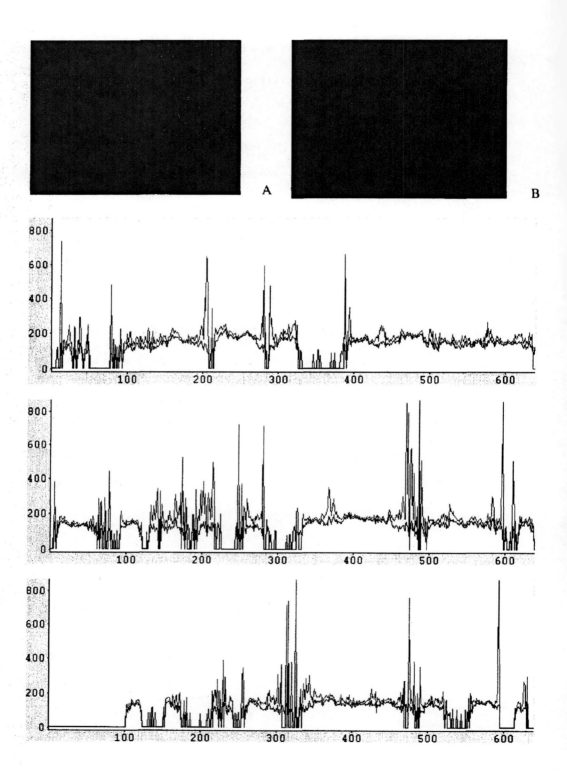

Figure 9: Comparison of adaptive (A) and baseline (B) approaches applied to images of RTDP-labeled living adenocarcinoma (SEG) cells. Line profiles are shown are shown for three lines as plots of lifetime (hundreds of ns, y-axis) and pixel number (x-axis), for both adaptive (green lines) and baseline (red lines) approaches.

The effect of both baseline and adaptive haze removal on a fluorescence lifetime image of living SEGs incubated with RTDP is shown in Fig. 9. Also illustrated are the lifetime values at three different horizontal profiles within the image. Several differences between the two methods are evident from the images. Firstly, haze and background removal is more effective for the adaptive approach. As a result spatial resolution is slightly better for the adaptive approach as well. Secondly, the lifetime profiles for the adaptive approach show significant spikes of inordinately high value and are likely inaccurate. Notably, the 'spiky' values are seen only on the boundaries of the fluorescent cells, something observable with other samples too (data not shown). The lifetime profiles for the baseline image, on the other hand, are relatively flat and require no filtering for spikes. Lastly, it is observable that lifetime values for the baseline approach are lower as compared to the adaptive approach. This is again consistently observed for any fluorescent sample, and is a property of the difference in analysis between the adaptive and baseline approaches.

For a more mathematical reasoning of these observed differences, we can use a simple 2-gate model of single-exponential decay (as described earlier), lifetime at any given pixel can be evaluated as:

$$\tau = \frac{t_2 - t_1}{\ln\left(C_1 / C_2\right)} \tag{7}$$

Where t_2 (second gate) > t_1 (first gate), and hence intensity counts $C_1 > C_2$. With adaptive haze removal, a variable intensity value (say x_1, x_2) that is is subtracted from every pixel:

$$\tau = \frac{t_2 - t_1}{\ln\left(\frac{C_1 - x_1}{C_2 - x_2}\right)} \tag{8}$$

Since $C_1 > C_2$ and $x_1 > x_2$, the ln term does not undergo significant change and lifetime information is generally conserved. With baseline haze removal, a fixed intensity value (say 'x') is removed from every pixel:

$$\tau = \frac{t_2 - t_1}{\ln\left(\frac{C_1 - x}{C_2 - x}\right)} \tag{9}$$

The denominator can be rewritten as:

$$= \ln\left[\left(\frac{C_1}{C_2}\right) * \frac{1 - \frac{x}{C_1}}{1 - \frac{x}{C_2}}\right] \tag{10}$$

$$= \ln\left[\left(\frac{C_1}{C_2}\right)\right] + \ln\left[\frac{1 - \frac{x}{C_1}}{1 - \frac{x}{C_2}}\right] \tag{11}$$

It follows that:

$$= \ln\left[\frac{1 - \frac{x}{C_1}}{1 - \frac{x}{C_2}}\right] > 0 \tag{12}$$

The additional positive term in the denominator results in a decrease in estimated lifetime, which explains why the baseline process consistently yields lower lifetime values than the adaptive approach. The additional term also likely explains the edge effect/spiky lifetime values; at low SNR, C_1 is only slightly higher than C_2, so C_1/C_2 is close (but never equal) to 1, hence $\ln(C_1/C_2)$ is close to 0. This explains the inordinately high lifetime values seen with the adaptive approach. Due to the additional $\ln[(1-x/C_1)/(1-x/C_2)]$ denominator term in the baseline approach, this effect is mitigated, at the cost of resolution and limited error in lifetime determination.

For all work reported in this dissertation, the adaptive noise removal approach was used. Spiky lifetime values were removed by setting the pixel lifetime value to zero. Given the flexibility of the imaging algorithm, it is quite possible to use alternate actions (e.g. set pixel value to image average or to a programmable maximum value).

Chapter 3 INTRACELLULAR OXYGEN SENSING IN LIVING CELLS

3.1 Introduction

Cancer mortality presents one of the leading medical challenges in the United States, accounting for almost 25% of all deaths [84]. While research has been extensive in this area, loss of life due to cancer was unchanged from 1950 to 2002 (193 deaths per 100,000), a trend explained to some extent by increased diagnosis and awareness in the general populace [84]. Cancer prevention through lifestyle changes and early detection is the most effective means of lowering these mortality rates.

Given the success of endoscopy as a minimally-invasive imaging modality for early cancer detection in the esophagus, the identification of an optically discernable mechanism that is consistently altered in a malignant process could prove very useful, not only for further developing therapeutic targets, but also for understanding disease pathogenesis and developing minimally invasive optical technologies for early detection. A recent clinical study presented evidence for the presence of detectable levels of NAD(P)H fluorescence in human epithelial tissues *in vivo* and demonstrated that NAD(P)H may be used as a quantitative fluorescence biomarker for *in vivo* detection of dysplasia in the esophagus [51]. Here, the underlying biological basis for endogenous fluorescence changes that occur during the course of esophageal cancer progression was investigated by employing a unique, time-resolved optical molecular imaging approach that provides a molecular snapshot of metabolic function in living human esophageal cancer cell models.

Knowing the complicated nature of cellular machinery, it is not surprising that pervasive changes in metabolic function almost always manifest as ill health at the anatomic level. The oxidative phosphorylation cycle is an attractive point of analysis in this regard: several key biomolecules such as NAD(P)H, $FADH_2$, oxygen, and ATP are consumed/generated during this process. Of these, NAD(P)H provides a bright, endogenous fluorescence signature that is detectable optically without the use of exogenous dyes [85]. Oxygen, while not fluorescent, can quench fluorescence in transition-metal complexes [11] and a class of ruthenium dyes has emerged as the biological probe of choice for oxygen sensing that has been studied for photochemistry, oxygen response, and toxicity [86, 87].

The large temporal dynamic range (600 ps – infinity) of FLIM are unique and essential for the purpose of this study: UV excitation and an ultrafast gated camera enable imaging of endogenous fluorophores such as NAD(P)H with sub-nanosecond lifetimes, while visible excitation from a low repetition rate laser source and large gates enable measuring lifetimes for ruthenium dyes, which are typically hundreds of nanoseconds. FLIM has been successfully applied to the study of NAD(P)H fluorescence, as well as lifetime modulation of oxygen sensitive ruthenium tris(2,2'-dipyridyl) dichloride hexahydrate

(RTDP), in living human bronchial epithelial cells [59, 70]. Here, the oxygen sensitivity of RTDP was calibrated to provide quantitative oxygen measurements in living cells under strict, temperature-controlled conditions using a fiber-optic oxygen sensor. Approaches for oxygen measurements in biological tissue have included, but are not limited to, electrochemical Clark-type electrodes (consume oxygen), NMR (requires high signal levels) and SECM (invasive, requires electrolyte media) [88, 89]. Fluorescence, on the other hand, is an established technique that meets the demanding criteria of high sensitivity and spatial resolution required for intracellular oxygen sensing. Oxygen affects biological systems at all strata of organization, from the subcellular (ATP production) to the cellular (proliferation) and supracellular level (eg. arterial oxygen), so an accurate and reliable method to quantitatively measure oxygen levels is critical for understanding as well as controlling systemic response.

The FLIM system was applied towards studying metabolic function in two related cell lines: normal human squamous esophageal epithelial cells (HET) and Barrett's adenocarcinoma esophageal cells (SEG). SEG were used as a model cell line for assessing the effect and capability of RTDP in the intracellular environment. Both cell lines were imaged for NAD(P)H and assessed for oxygen levels. Measurement of multiple components of an ordered pathway such as oxidative phosphorylation not only provide a more complete perspective on cancer progression, but also offer endogenous targets for clinical optical diagnostic technologies for cancer prevention, without the need for exogenous contrast agents.

3.2 Materials, Instrumentation, and Methods

3.2.1 Fluorescence Lifetime Imaging Microscope (FLIM)

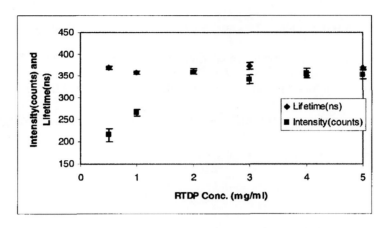

A

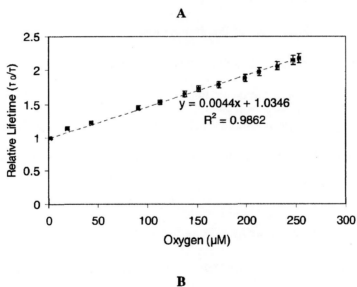

B

Figure 10: A) Plot of RTDP intensity (pink squares) and lifetime (blue diamonds) vs. RTDP concentration. Lifetime was contant over a large range, while intensity varied with concentration before saturating. B) RTDP Calibration Curve: Plot of relative lifetime (τ_0/τ) vs. oxygen levels (μM). RTDP calibration indicated a linear relationship between oxygen levels and relative lifetime, which was in good agreement with the Stern-Volmer equation. Over multiple runs $K_q = 4.5 \pm 0.4 \times 10^{-3}$ μM^{-1}. The intercept $\neq$ 1, indicating some degree of experimental variance. The calibration could differentiate between oxygen levels differing by as little as 8μM.

The FLIM system has been described in Chapter 2. For RTDP experiments, a dichroic mirror (Q495lp, Chroma Technology Corp., Brattleboro, VT) delivered the beam into a 40× Fluar (1.3NA, Zeiss, Jena, Germany) objective for sample excitation with an illumination diameter of 150 μm. Gate widths were controlled by slaving the ICCD gate to an external logic signal. Intensifier (MCP) voltage was set at 700 V for the RTDP solution and 800 V for cell samples. For imaging cellular NADH, the ICCD was operated in Comb mode and the dichroic mirror (350lpDC Chroma Technology Corp.) was changed accordingly.

3.2.2 Image Analysis

Image acquisition and processing for FLIM was described in Chapter 2. For RTDP imaging, the linear scale of the lifetime map was converted to a binary, two-color scale for the sole purpose of easier visualization. The oxygen distribution image was generated by directly applying the calibration (described below) to the lifetime map without pre-processing. Finally, threshold filters were applied that set pixels with unlikely oxygen levels (both low and high) to zero. This is reasonable given the range of physiological oxygen values [90].

3.2.3 Temperature Control

For temperature control, the Delta T controlled culture dish system (Bioptechs, Butler, PA) was implemented along with the Delta T perfused heated lid (Bioptechs) and was used for all experiments, unless noted otherwise. The heated lid had two ports which could be used for perfusion purposes. A remote controller was used to adjust temperature settings. For closed chamber experiments the FCS2 chamber system and controller

(Bioptechs) was used, where cells were enclosed between a cell plate and a perfusable gasket with a media holding capacity of 1 ml. Due to the large thermal mass of the objective, a separate Objective Heater (Bioptechs) with its own controller was also installed. All temperature control systems had an operating range from ambient to 45°C with accuracy within ±0.2°C.

3.2.4 Confocal Microscopy

An independent Zeiss LSM 510 microscope (Zeiss, Jena, Germany) was used to acquire confocal NADH fluorescence images via 364 nm excitation (emission: 435 nm-485 nm). Images were acquired via LSM 5 Software Release 3.2 (Zeiss) in real time. Detector gain and offset were configurable via the software interface and were adjusted to avoid detector saturation. Offline image analysis was done using LSM Image Browser 3.2 (Zeiss). The only image parameters adjusted for enhancement were brightness and contrast and these changes were uniformly applied across the entire image. A range indicator function in the software provided information on color saturation (via brightness enhancement) or zero pixel values (via increased contrast). The function used a pseudo-color scheme where all saturated pixels were colored red, all zero pixels were colored blue and all other were plotted on a linear grayscale. Adjusting brightness and contrast while maintaining the entire image within the grayscale range allowed preservation of image information while avoiding loss or generation of data.

3.2.5 RTDP Characterization

0.025 g of RTDP (ruthenium tris(2,2'-dipyridyl) dichloride hexahydrate) powder obtained from Sigma-Aldrich (#224758, St. Louis, MO) was dissolved in 25 ml phosphate buffer saline (PBS, Invitrogen, Carlsbad, CA) to create a stock solution of 1 mg/ml, or 1.34 mM. Absorption spectra of RTDP were obtained from a Beckman-Coulter DU-800 Spectrophotometer (Ontario, Canada) over a range of 200-800 nm. The fluorescence emission spectrum was extracted using a Jobin-Yvon Fluorolog-3 (Edison, NJ) in a range of 300-700 nm. Temperature response of RTDP lifetime was studied using 2 ml of stock solution and FLIM. Initially, ambient temperature was 22°C and the solution was exposed to normal atmospheric conditions. Temperature was varied in the range of 25-45°C in steps of 2.5°C. At each temperature point, lifetime images were acquired three times and averaged. Multiple runs were used to validate the slope of the plot obtained. FLIM was used to study RTDP lifetime independence of its concentration. Six different RTDP concentrations in the range 0.5-5 mg/ml were used for intensity and lifetime studies (Fig. 10A). The temperature was set to 37°C (physiologic value), gate width at 50 ns and the HRI gain at 700 V. The settings were selected to maximize signal at the CCD without saturation when the highest concentration of 5 mg/ml was measured for lifetime. Below 0.5 mg/ml, the intensity was too weak to reliably evaluate lifetimes without changing gain settings to manipulate intensity. Three readings were obtained per concentration and were averaged for intensity and lifetime.

3.2.6 Calibration of Oxygen Sensitivity of RTDP

Independent oxygen measurements for RTDP calibration were performed using a commercial fiber-optic oxygen sensing system (FOXY, Ocean Optics, Dunedin, FL). The FOXY system is a fluorescence-based, spectrometer-coupled chemical sensor used for spectral analysis of both dissolved and gaseous oxygen with a manufacturer specified range of 0-40 ppm (>1 mM) for oxygen dissolved in water, a resolution of 0.02 ppm, and response time < 1 sec. The FOXY was calibrated using a two-point linear fit. The first point was obtained by immersing the FOXY probe in a solution of sodium hydrosulfite (dithionite) dissolved in DI water through which nitrogen had been bubbled, yielding a solution with near-zero oxygen concentration (sodium dithionite is a potent oxygen scavenger and bubbling nitrogen removed virtually any traces left). The second point was obtained by immersing the FOXY probe in a solution of DI water, which has a known oxygen concentration of 7.1 ppm or 222 μM at 37°C under normal atmospheric O_2. Data from these two points was used to calibrate the FOXY via a software interface.

Next, a lifetime-oxygen calibration of RTDP was performed on the stock solution using FLIM. Gas tubing was attached to one port on the heated, perfused lid while the FOXY was inserted into the solution via the other port. Under equilibrium conditions, the oxygen level was similar to the calibration value, i.e., 7.1 ppm. Nitrogen was flowed over the sample and a drop in oxygen was recorded via FOXY. Once the FOXY output stabilized at <0.1 ppm, the lifetime was recorded, the flow of nitrogen was stopped and the solution was allowed to recover by equilibrating with atmospheric oxygen. During this process, several concurrent lifetime and oxygen measurements were taken.

Once the solution reached equilibrium, the solution was perfused with oxygen and an increase in levels was correspondingly observed by the FOXY. The oxygen flow was then cut-off and while the solution stabilized toward equilibrium, readings were taken in a manner similar to that described above. The data was verified by repeating the calibration multiple times. Average values were used for cell studies.

Data analysis - Stern-Volmer Equation: Oxygen is one of the best known collisional quenchers and its effect on RTDP can be effectively described by using the Stern-Volmer Equation [11]. Since increased temperatures lead to faster diffusion for both fluorophore and quencher and hence higher collision quenching rates, it is characteristic for K_q to increase with temperature [11]. To evaluate K_q, the highest lifetime value attained (when $[O_2]$ <0.1 ppm) was set equal to τ_0. This was reasonable, since the FOXY had a minimum detection limit of 0.02 ppm and it is near impossible to completely deplete oxygen. For all other values, τ_0/τ was evaluated and plotted as a function of $[O_2]$. Line fitting of all data points and regression analysis was carried out using statistical software. The slope of the line yielded a single value of K_q. Over multiple runs, K_q was averaged to provide a mean value that was used for analyzing cell lifetimes (see below). A key advantage of the relative lifetime, Stern-Volmer approach was that any variability in RTDP lifetime associated with aging or other ambient conditions was systematically eliminated by determining τ_0 from the known (stock) solution for each experiment.

3.2.7 Cell Preparation

The HET-1 is a human squamous epithelial cell line immortalized by transfection of the SV40 T antigen early region gene [91]. The SEG-1 (Barrett's) adenocarcinoma esophageal cell line was derived from Barrett's-associated adenocarcinoma of the distal esophagus [92]. HET and SEG were propagated in 100 cm^2 culture flasks in DMEM media (Invitrogen) containing 10% fetal bovine serum (FBS) and cultured at 37°C under standard conditions. Upon reaching 70% confluence, the cell cultures were split using trypsin and divided at a ratio of 1:4. For open-air studies, the cells were plated a day before experimentation. About 6 hours before analysis, 1 ml of RTDP stock solution was added to the cells. The cells were washed with PBS or media (without phenol red) before being transferred to the microscope stage. This process was mirrored for both cell lines whenever comparative studies were done. Cell viability tests (Trypan Blue) yielded >85% viability after 6 hrs of incubation with RTDP, indicating that significant cytotoxicity was not occurring within the incubation period. No staining was necessary for NADH studies.

For closed chamber RTDP studies, the cells were plated a day before on cell plates (see Temperature Control) and the RTDP stock solution was added about 6 hours before experimentation. After washing with PBS, the plates were assembled into the closed chamber system, sealed shut with media and mounted onto the microscope stage.

3.2.8 RTDP Lifetime Measurement in Cells

All controllers were set at 37°C after the cells were loaded onto the stage. The objective was used to focus on an appropriate region with the cells clearly visible and lifetime measurements were taken. Each measurement was repeated multiple times. For time-lapse experiments, measurements were taken periodically. Lastly, for each experiment, the lifetime of the stock solution was also obtained. The calibration (K_q) was applied to determine intracellular oxygen concentration. Values of lifetime (τ) was evaluated for the stock solution of known oxygen level and fed into the Stern-Volmer equation to obtain τ_0. A software macro was implemented to generate oxygen distribution images based on the calibration data and using the lifetime images as input. The oxygen distribution was evaluated on a pixel by pixel basis for each lifetime image by feeding back the value of τ into the Stern-Volmer equation. For each cell visible in the lifetime image, τ_{cell} was evaluated as the average of all pixels comprising the image of the cell. Hence, oxygen levels were assumed to be uniform within a cell. Average intracellular oxygen levels were evaluated from the oxygen distribution in a similar manner. For each experiment, oxygen levels from at least 5 cells were averaged to obtain mean values.

3.3 Results

3.3.1 NADH Measurements in HET and SEG

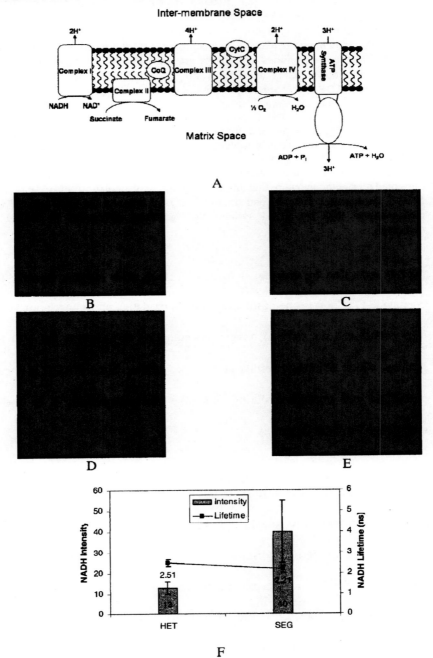

Figure 11: A) Oxidative phosphorylation in mitochondria. Complex I (NADH dehydrogenase) converts NADH to NAD^+ and passes on electrons to carrier CoenzymeQ (CoQ) while pumping hydrogen ions into the intermembrane space. Complex II (succinate dehydrogenase) can also generate and pass on electrons to CoQ via a complex internal mechanism which is initiated by conversion of succinate to fumarate, a step in the Krebs cycle. CoQ transfers the electrons to an intermediary complex III (CoQ – cytochrome C oxidoreductase), which in turns enhances the proton gradient by pumping hydrogen ions against the gradient. The electrons are transferred to complex IV (cytochrome oxidase) via cytochrome C (CytC) which in turn hydrolyses oxygen to water and also pumps protons against the gradient. The ATP synthase complex (complex V) moves protons down the gradient, converting osmotic energy to chemical energy via ATP synthesis from ADP in a process known as chemiosmotic coupling. A more detailed explanation of the process and the unique structure of ATP synthase can be found in any standard biochemistry text. B) NADH fluorescence and C) Mitotracker-stained SEG images. For the Mitotracker, excitation = 543 nm and emission = 636 nm. Mitotracker Red is a commercially available stain used for tracking mitochondria within living cells. Fluorescent signals from both markers were found to co-localize. Confocal intensity images of NADH fluorescence in HET (D) and SEG (E): Illustration of differences observed with the Zeiss 5 LSM. The SEG consistently presented a brighter signature than the HET by approximately 2.5-fold. F) Plot of differences in NADH fluorescence intensity and lifetime observed between the HET and SEG over multiple measurements with the FLIM system. No significant differences in NADH lifetime were observed.

NADH reduction by complex I to NAD^+ along with electron transfer to the carrier coenzyme-Q (CoQ) is one of the first steps in oxidative phosphorylation (Fig. 11A). This step contributes not only to oxygen consumption downstream but also to the proton gradient which ultimately results in ATP production. Since NADH is generated during glycolysis and consumed during oxidative phosphorylation, it is one measure of completion for these processes.

The term NAD(P)H is used to point out the spectral similarity of NADH and NADPH, both of which have a wide excitation range centered at approximately 350 nm and an emission peak at 460 nm. Since other endogenous fluorophores have overlapping excitation in the UV range, preliminary studies were conducted to observe the sub-cellular origin of fluorescence. Fig. 11 presents confocal fluorescence intensity images of SEG incubated with a commercial mitochondrial stain (Mitotracker Red). Fig. 11B shows that NAD(P)H fluorescence clearly has the same origin as the Mitotracker fluorescence (Fig. 11C), indicating that fluorescence observed arose mostly from mitochondrial NADH and not NADPH.

Figs. 11D, 11E provide confocal NADH fluorescence intensity images from HET and SEG, respectively, under identical measurement settings. It is visually evident that the SEG have a starkly brighter fluorescence signature than the HET. Similar results were also obtained with FLIM and overall the SEG exhibited 2-5 fold higher fluorescence signals than the HET. This difference in signal from NADH was observed to be statistically significant ($p<0.05$, Fig. 11F).

FLIM measurements were made for NADH lifetime in both cell lines. Values obtained were $\tau_{HET} = 2.51 \pm 0.16$ ns and $\tau_{SEG} = 2.21 \pm 0.16$ ns and were within the range of values reported in literature [93]. The differences in lifetimes were not statistically significant, therefore it can be established that the fluorescence intensity differences between the HET and the SEG were mainly attributable to differences in intracellular NADH levels. The observation reported here that esophageal cancer cells have higher absolute NADH levels than normal esophageal cells could explain the higher levels of NAD(P)H reported in dysplastic vs. nondysplastic esophageal tissues via clinical studies described earlier (see introduction) [51].

3.3.2 RTDP Calibration

RTDP has been studied for its photophysical and photochemical properties and recently found application in biomedical research as a molecular oxygen sensor [86]. With a long fluorescence lifetime (hundreds of nanoseconds), easy uptake by cells, and minimal cytotoxic and phototoxic effects, RTDP provides a nearly ideal means of assessing intracellular oxygen.

Measurements made with the FLIM system indicated that RTDP lifetime decreased with increasing temperature and that this decrease was almost linear:

$$\tau = -4.4975*T + K, \; r^2 = 0.9867 \tag{13}$$

where τ = lifetime and T = temperature in degree Celsius and K was a constant, emphasizing the need for temperature control in lifetime studies. The slope was in good agreement with previous work, which measured RTDP lifetimes within a similar temperature range [87].

RTDP calibration results at 37°C indicated a linear relationship between oxygen levels and relative lifetime, which was in good agreement with the Stern-Volmer equation (Fig. 10B):

$$\tau_0/\tau_x = 1 + K_q[O_2]_x \tag{14}$$

where K_q is the Stern-Volmer quenching constant. Relative lifetime at a given oxygen level $[O_2]_x$ was evaluated as the ratio of uninhibited RTDP lifetime (i.e. 0% oxygen, or τ_0) to τ_x. Over multiple runs, the value of K_q was evaluated to be $K_q = 4.5 \pm 0.4 \times 10^{-3} \, \mu M^{-1}$. This is higher than other reported values which were measured at room temperature, confirming that K_q increases with temperature [90]. System resolution was determined by a) the resolution of the oxygen sensor used for calibration (±0.6 µM) (see Methods section) and b) the lifetime variance of the FLIM system (±2% of lifetime, or ±6 ns for

RTDP). Using these values in Eq. 14, it was estimated that the calibration could reliably differentiate between oxygen levels differing by as little as 8 µM.

3.3.3 Oxygen Measurements in HET and SEG

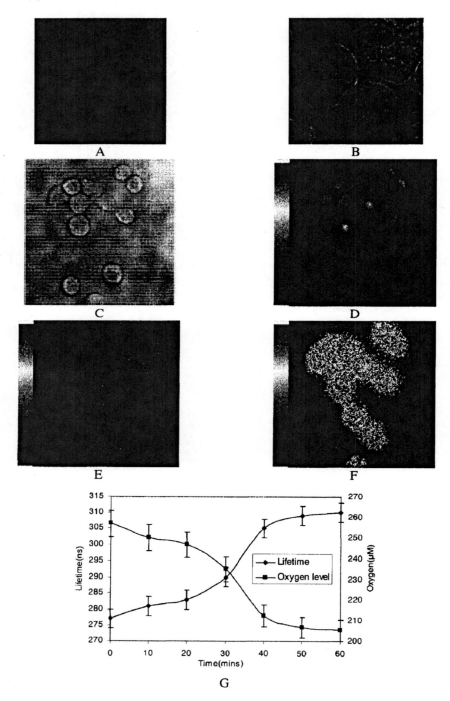

Figure 12: (A) confocal fluorescence and (B) Differential Interference Contrast, or DIC images of SEG. SEG were incubated with the RTDP dye prior to imaging (see Methods section). FLIM images of SEG incubated with RTDP: (C) DIC, (D) fluorescence intensity (0-400 counts), (E) lifetime (0-1000 ns) and (F) oxygen (0-500 µM). Note that one cell in the bottom of (C) shifted position in (D) in the time lapse between these two images. (G) The results of depletion experiments on SEG (see Methods section). Cellular viability is compromised with the passage of time and this results in the lifetime / oxygen content leveling off towards the end.

Confocal measurements indicated that RTDP distributed uniformly in the cytoplasm without any evident aggregation or adherence to organelles (Fig. 12A,B). It yielded a strong intracellular luminescent signal without visible cytotoxic effects or membrane damage. Similar results were obtained with the FLIM system for both HET and SEG with excellent cell viability (>85%).

Fig. 12C-F presents representative images of SEG cell lines after analysis for lifetime and oxygen levels. The cell positions (C) overlaps with RTDP fluorescence (D). Note that the fluorescence intensity image (D) is non-uniform, indicating that the cells differentially absorb RTDP. The lifetime map (E) generated from the intensity map (D) indicated a clear uniformity across different cells, signifying similar oxygen levels and reflecting intensity independence of lifetime. RTDP calibration was used to generate an oxygen distribution map (F) that implied uniform oxygen levels of approximately 285 µM. The oxygen levels were found to be $[O_2]_{SEG}$ = 260.16 ± 17 µM. While higher than reported *in vivo* values, these levels were comparable to values generally observed in living biological samples during *ex vivo* measurements. This is a leap forward from measurements of extracellular oxygen alone, which assume linear correlation with intracellular oxygen. Gradients between intracellular compartments and extracellular space are known to exist but cannot be explained by simple diffusion rates alone [59, 90].

Time-lapse FLIM measurements were made on SEG under airtight conditions to verify that decrease in oxygen levels due to cellular consumption would be reflected in RTDP lifetime increase. The results are plotted in Fig. 12G. As expected, RTDP lifetime increased by 35 ns over the course of an hour, reflecting a decrease in oxygen levels ($\Delta \approx -50$ μM) due to gradual consumption by cells. These experiments demonstrate the sensitivity of FLIM for oxygen sensing in living cells.

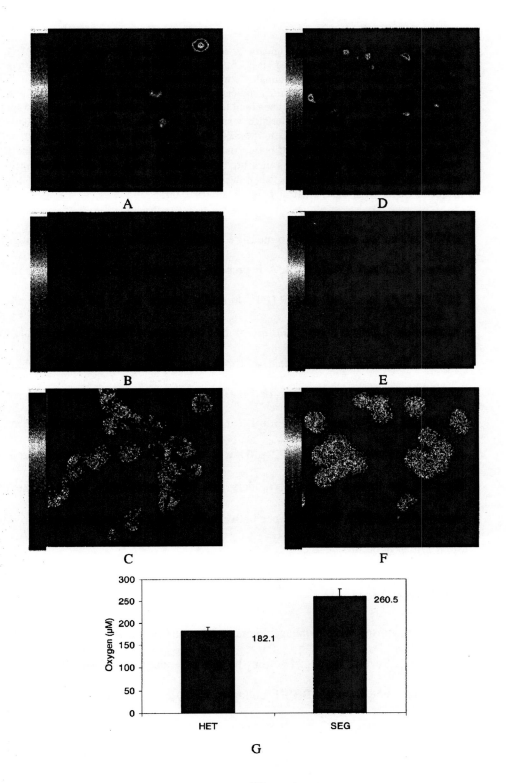

Figure 13: RTDP fluorescence intensity (0-600 count) (A,B), lifetime (0-1000 ns) (C,D) and oxygen (0 – 500 µM) (E,F) maps of HET (A,C,E) and SEG (B,D,F). The intensity images (A,B) could not be reliably used to discriminate between the two different cell lines. The binary lifetime maps (C,D), on the other hand, plainly indicate different lifetimes for these two cellular species, with the SEG recording lower lifetimes than the HET. For the given case, τ_{HET} = 225 ns and τ_{SEG} = 170 ns. The mean lifetime difference was found to be Δ_τ = 44 ± 7.48 ns. Logically, this translated into higher oxygen levels in the SEG vs. the HET using the calibration derived earlier, as can be seen in the oxygen distribution maps (C,F). G) illustrates the differences in oxygen levels within the HET and SEG as measured over multiple runs and assessed using the RTDP calibration. The mean difference between the two cell lines was hence evaluated as $\Delta_{[O]2}$ = 78 ± 13 µM and this value was statistically significant (p< 0.001).

RTDP calibration was applied to measure quantitative differences in intracellular oxygen between HET and SEG. Fig. 13A-F presents comparative images of HET (A,C,E) and SEG (B,D,F) incubated with RTDP. Intensity images (A,B) do not yield any useful information. Lifetime maps (C,D) are plotted on the same bicolor lifetime scale to better illustrate disparities: the SEG clearly exhibit lower lifetimes than the HET, which in turn translates to higher oxygen levels (E,F). This trend was consistent, with $[O_2]_{HET}$ = 182.08 ± 9.38 µM and $[O_2]_{SEG}$ = 260.16 ± 17 µM (Fig. 13G). The difference between $[O_2]_{HET}$ and the extracellular media (222 µM) was within the range of gradients reported in literature for various other cell models [94, 95]. Interestingly, the opposite trend is observed in the SEG. The difference between HET and SEG for the images provided was approximately 60 µM.

3.4 Discussion

Patients diagnosed with esophageal cancer have one of the lowest 5-year survival rates (14%) [84]. Several reports describe the use of endogenous tissue fluorescence (e.g., autofluorescence from NAD(P)H, collagen, FAD) as a biomarker for dysplasia in the esophagus, and recent studies have corrected for signal distortions from other sources (e.g., hemoglobin absorption) to yield quantitative trends (differences) between fluorescence signatures of NAD(P)H and collagen in normal and dysplastic tissue *in vivo* [51, 96, 97]. Thus, it was possible to observe consistent changes in absolute NAD(P)H and collagen fluorescence levels of multiple patients and accurately correlate them with the degree of neoplastic change [51]. A key strength of tissue spectroscopy is the acquisition of data in the appropriate local environment. Although cellular NAD(P)H is a marker of esophageal cancer progression, the exact mechanisms are unknown in part due to the poor spatial resolution of fiber-optic based spectroscopic measurements. Thus, spectroscopy is unable to distinguish between cellular NADH in mitochondria and NADPH in the cytoplasm, due to their spectral similarity (hence the term NAD(P)H). Both these components, however, have contrary roles: NADH takes part in the electron transport chain in mitochondria to generate ATP, while NADPH is the reducing agent for anabolic reactions such as fatty acid synthesis that consume ATP. This article reports results from a dual confocal-FLIM approach towards measuring NAD(P)H fluorescence in cell models characteristic of normal and adenocarcinoma esophageal state (HET and SEG, respectively). High resolution confocal images of NAD(P)H and Mitotracker fluorescence (Fig. 13B,C) provided sub-cellular imaging of co-localization of these two signals, indicating that endogenous fluorescence arose primarily from mitochondria and was NADH. Using FLIM, differences in NADH fluorescence intensity were observed between HET and SEG, while there was no significant variation in NADH lifetime between cell lines (Fig. 13F), indicating that differences in fluorescence intensity were mainly attributed to differences in NADH concentration, and did not arise from fluorescence quenching. Intensity results were corroborated with the confocal system

(Fig. 13D,E) and confirmed that SEG contained significantly higher NADH levels than the HET, a trend similar to and supporting clinical observations.

Elevated NADH levels in SEG raises the question whether it is being excessively produced during glycolysis or consumed less due to possible mitochondrial dysfunction. Mitochondria provide a more logical starting point for further analysis: they are well characterized, complex organelles which play key roles in several cellular functions such as fatty acid metabolism, calcium homeostasis and most importantly, ATP production via the oxidative phosphorylation (OXPHOS) pathway (Fig. 13A) [98]. It is clear that defects/alterations in OXPHOS enzymes have links to some forms of cancer, neurogenerative diseases, diabetes and aging [98, 99]. The first step in OXPHOS involves the reduction of naturally fluorescent NADH to its non-fluorescent analog NAD^+ via the membrane-bound complex I enzyme. Given the sequential nature of OXPHOS, it is logical that variations in NADH will affect the rate of downstream processes; i.e., consumption of oxygen and generation of ATP. ATP is the energy currency of the cell but is not an intrinsic fluorophore and commercial fluorescence indicators for ATP do no differentiate between ADP and ATP, making analysis difficult [100]. Oxygen, on the other hand, is likely the most pertinent indicator of completion of cellular respiration, in general, and OXPHOS, ATP production, in particular. Experiments conducted on HET and SEG indicated the potential of RTDP as an oxygen sensitive ruthenium dye that was well tolerated by living cells and yields a bright fluorescence signal to provide a molecular capability for analyzing intracellular oxygen levels in these cell models (Fig. 13A-F). After calibrating the oxygen sensitivity of RTDP lifetime via FLIM, modulation of intracellular RTDP lifetime was demonstrated via the gradual depletion of oxygen supply to SEG (Fig. 13G). Using FLIM, it was estimated that $[O_2]_{SEG} = 260.16 \pm 17$ μM under cell culture conditions. Oxygen levels in HET were lower than SEG, observed as $[O_2]_{HET} = 182.08 \pm 9.38$ μM, bearing an average difference $\Delta[O_2]_{SEG-HET} \approx 80$ μM (Fig. 13A-G). This difference was statistically significant ($p<0.05$). It is important to note that cells in culture are exposed to higher oxygen levels than *in vivo* (especially for cancer cells which normally inhabit a hypoxic environment *in vivo*) and their behavior under these conditions can be meaningfully interpreted to understand causal behavior. This comparative tactic of tracking differences between normal and altered cell lines is easily extensible to other disease models and/or analytes other than NADH and oxygen.

The phototoxicity of RTDP in living cells was extensively documented previously [86]. The protocol used in that study included supplementation of the culture media by a solution of RTDP immediately prior to 457 nm excitation in a Bio-Rad MRC1024 laser scanning confocal system. Based on the data provided, we estimate that each illumination delivered approximately 0.8 J/cm^2 of energy on the sample. Depending on the concentration of RTDP used, photodamage to living cells occurred when the cumulative light dose exceeded 20-80 J/cm^2 (2-0.2 mM). In contrast, the protocol employed here involved incubating the cells with 1.3 mM RTDP stock solution for 6 hours and washing away the dye with PBS prior to experimentation. The intracellular concentration of RTDP, while not measurable, was clearly lower than the stock solution as inferred from intensity based fluorescence measurements. For the study described here, the laser pulse energy at the sample was 15 µJ and the illumination area was approximately 0.5 mm in diameter. Hence, the energy per unit area was 1.9 mJ/cm^2 for a single measurement. Further, the lifetime analysis protocol required intensity images captured at six different delays, at five measurements per delay. Therefore, it required thirty measurements to deduce lifetime, resulting in a total energy exposure of 0.057 J/cm^2 at the sample. Lastly, six lifetime measurements were taken for oxygen depletion experiments and the cumulative energy deposited was 0.34 J/cm^2. This is significantly lower than the threshold observed previously [86]. Coupled with the extremely low levels of RTDP being excited, we reasonably infer that the FLIM-based experiments presented here do not cause photodamage.

The studies presented here confirm that optical measurements of fluorescence from endogenous mitochondrial NADH might be useful for non-invasively detecting altered metabolic function in esophageal cancers *in vivo*. Use of exogenous agents such as RTDP, provide a complementary, highly specific approach for quantitative, intracellular oxygen sensing in living cells. Results presented here indicate that a possible alteration in cellular metabolism in SEG early in the OXPHOS pathway could cause decreased NADH consumption (higher intracellular levels), which would then lead to lower oxygen consumption downstream and hence elevated intracellular oxygen levels relative to HET, as observed. While reports on basal cell carcinoma indicate that OXPHOS complex I gene and/or protein modification might result in impaired activity, no such report was found for esophageal cancers [101].

The flexibility of FLIM to detect endogenous cellular fluorescence from NADH, as well as exogenous fluorescent tags and labels, makes the method compatible with endoscopic clinical imaging studies in living human tissues. A large temporal dynamic range and a microscopy imaging modality provide the ability of measuring both fluorescence and phosphorescence with subcellular spatial resolution. High temporal resolution and single-shot imaging aid in resolving molecular events such as conformational and viscosity changes. Temperature and perfusion control of the cellular environment are beneficial for viable cell imaging under physiologic conditions and to observe cellular response to conditions such as hypoxia and heat shock. Further, thick biological specimens can be imaged in layers by rejection of out of focus light (optical sectioning) via structured illumination to yield 3-D distribution of analytes (e.g., oxygen gradients in thick tissues). The oxygen sensing capability of FLIM was recently exploited to provide one of the first few reports on oxygen gradients in microfluidic bioreactors containing living cells (submitted).

The approach described here employed time-resolved optical imaging to probe metabolic pathways in living cells using endogenous and/or exogenous fluorescence agents. A calibrated approach for quantitatively estimating oxygen levels in HET and SEG using a tunable, temperature-controlled FLIM was presented, with the potential to perform absolute oxygen measurements in living biological systems with a resolution of 8 μM over the entire physiological range (0-300 μM). Previous studies provided a calibration curve for RTDP dissolved in growth buffer, but measurements were performed at room temperature and were made using electrodes, which consumed oxygen and required constant stirring of the media [90, 102]. Given the temperature-dependence of lifetime, oxygen solubility, and cellular biochemical response, a controlled study like this provides more accurate results. Precise measurements of intracellular oxygen provide important information in areas such as pharmaceutical research or systems biology and may aid in a better understanding of systemic disorders and their control.

Chapter 4 CALIBRATION AND VALIDATION OF INTRACELLULAR OXYGEN MEASUREMENTS

4.1 Introduction

Oxygen is the terminal electron acceptor of oxidative phosphorylation in the mitochondria and a key indicator of metabolic function in aerobic cells and organisms. Information obtained from accurate oxygen estimation has been applied for studying mitochondrial function, signaling pathways, effects of various stimuli, membrane permeability, differentiation between normal and disease, or to screen for new drugs [94, 103-105]. Various methods exist for estimating oxygen at different scales, whether in whole organs (MRI, NMR), in the blood (absorption), in tissues (electrochemical microelectrodes) or in individual cells (phosphorescence, fluorescence, EPR) [54, 59, 106, 107]. Optical approaches such as fluorescence microscopy provide the most viable approach for intracellular oxygen sensing with high sensitivity and spatial resolution. Fluorescence lifetime-based methods additionally circumvent most issues associated with intensity-based intracellular studies, such as photobleaching, scattering, variable uptake by cells, etc [70].

Chapter 3 presented a FLIM-based approach for intracellular oxygen sensing using the oxygen-sensitive dye RTDP. One concern that consistently arises is the validity of applying an oxygen calibration derived in PBS solution to measurements made in living cells, including how other intracellular factors might affect RTDP lifetime. Oxygen sensitivity of RTDP is widely accepted and exploited [87]. It has been demonstrated that RTDP lifetime is unaffected by solvent pH and ionic concentration (including Na^+, Ca^{2+}, Cl^-) within physiologic ranges, but is dependent on temperature, which is controlled by the FLIM approach. Solvent viscosity is also known to affect fluorophore lifetime, not only by changing the microenvironment but also by altering diffusion rates of both fluorophore and quencher. In one study, RTDP lifetime changes were reported over a range of solvent viscosities that is extreme in comparison to that encountered in biological systems (liquid media vs. rigid glass). Combined with previous studies, results presented here indicate that it is unlikely for viscosity to vary between HET and SEG or between cell cytoplasm and culture media/PBS. One question that remains unanswered is how intracellular biochemistry might affect RTDP lifetime, if at all, and provides the impetus for this study.

Irrespective of origin of the oxygen-dependent signal (i.e. for any oxygen-sensitive probe), a robust calibration method is necessary for quantification. Since excited-state fluorophore intensity and lifetime is inherently dependent on the microenvironment, a simple solution is to reproduce the intracellular molecular environment as the calibration media. Several kinds of solvents have been used, including PBS and varied culture media [36, 108]. While convenient, such media does not approach the molecular complexity of intracellular cytosol. Other work involved 'poisoning' the fluorophore-incubated cells to prevent oxygen consumption, followed by measuring variation of fluorescent signal with change in extracellular oxygenation [8]. While such an approach is more relevant, working with poisoned cells raises concerns about cellular response to hypoxia at very low oxygen levels, which could potentially affect the calibration curve. Hence, there is a need for a more accurate protocol for estimating intracellular oxygen in such *in vitro* samples.

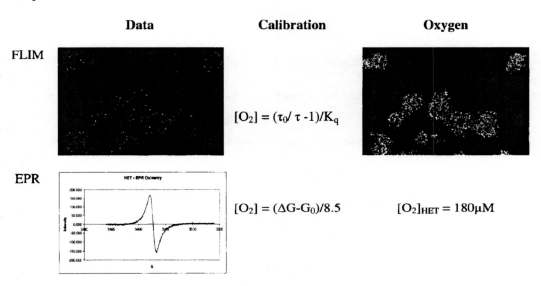

Figure 14: Illustration of FLIM (imaging) and EPR (Spectroscopy) methods for quantitative oxygen sensing in living cells. For FLIM, RTDP lifetime images serve as the raw data. Oxygen sensitivity of RTDP can be calibrated via the Stern-Volmer equation and applied to every pixel in the lifetime image to yield an oxygen distribution image.
EPR methods yield a spectral response of signal intensity vs. magnetic field strength. The peak-to-peak separation (designated by ΔG) of the EPR probe is oxygen sensitive and can be calibrated to provide a single oxygen value for the entire sample. This value for HET is comparable to that obtained via FLIM.

In this work, for the first time, lysate-FLIM studies were performed to refine the calibration derived earlier to make it sample specific. The revised calibration was then used to correct intracellular oxygen measurements that were previously reported [36]. Further, all fluorescence oxygen measurements were compared by complementary measurements via 'the gold standard' EPR (Electron Paramagnetic Resonance) oximetry technique. We propose a protocol for accurate *in vitro* oxygen estimation in living cells. An illustration of quantitative oxygen sensing via the FLIM and 'gold-standard' EPR Oximetry methods is shown in Fig. 14.

4.2 Materials and Methods

4.2.1 FLIM and Oxygen Sensing.

The FLIM system and RTDP calibration have been described previously [36, 59, 70].

4.2.2. Cellular Lysate Generation and FLIM Analysis

Cellular lysate of the HET-1 (human squamous esophageal epithelial cells) and SEG-1 (Barrett's adenocarcinoma esophageal cells) were generated using the NP-40 buffer and protocol. NP-40 was thawed on ice and 10μL/ml of Halt Protease Inhibitor was added for inhibition of serine-, cysteine-, metallo-, and aspartic acid-proteases in addition to aminopeptidases. 50μL/ml each of DNAse and RNAse was added to reduce viscosity due to DNA/RNA release.

Cells grown in culture flasks were scraped off into PBS, centrifuged and washed twice with ice-cold PBS. The NP-40 formulation was added in a ratio of 1ml buffer / 10^8 cells. The solution was placed on ice and vortexed for 30 seconds every 10 minutes. After 30 minutes, the solution was transferred to microcentrifuge tubes and run at 13000RPM for 10 minutes at 4°C. This allows the heavier debris (e.g. cell membrane components) to sediment, and the clear aliquot was removed and frozen at -80C. For lysate-FLIM experiments, 0.8ml of lysate was combined with 0.2ml of 50mg/ml RTDP in a 37°C heated dish, and lifetime measurements were immediately taken.

4.2.3 EPR Oximetry

EPR Oximetry is a spectroscopic technique to detect materials with unpaired electrons, such as free radicals and transition metal ions. It is considered the electron equivalent of NMR (nuclear magnetic resonance), except that it utilizes weaker magnetic fields at higher frequencies. The theory of EPR is as follows: all electrons have a magnetic moment, and can hence align along a magnetic field. Parallel alignment is the lower, stable energy state, while anti-parallel alignment is the higher, more unstable state. The magnetic field can cause the electron to oscillate between these two states upon absorption, yielding a bump in the absorption spectrum (Fig. 15). The extent (i.e. intensity) of the absorption peak as well as its width are informative about quenching of absorption, which results in broadening of the spectrum. The derivation of the absorption spectrum provides two parameters of note: peak intensity, and peak-to-peak width. Spectral line broadening then affects the peak-to-peak value.

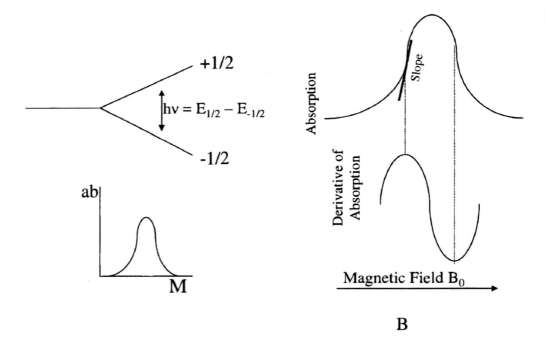

Figure 15. Illustration of EPR theory and operation. A) Energy absorption by the electron to shift between parallel (-1/2) and anti-parallel (+1/2) states results in a peak in the absorption spectrum, denoted as a plot of absorption (ab) vs. Magnetic Field (M). B) The derivation of the absorption spectrum is a measure of the slope and indicative of the environment of the spin probe.

Even though molecular oxygen is paramagnetic, having a triplet ground state, no EPR spectrum has been recorded for oxygen dissolved in liquids near physiological temperatures. Fortunately, EPR methods exist where bimolecular collisions of molecular oxygen with spin-label probes are used to monitor oxygen levels, that can be calibrated for purposes of quantitative estimation.

The oxygen sensitive spin probe used, lithium 5,9,14,18,23,27,32,36-octa-n-butoxy-naphthalocyanine (LiNc-BuO), has been described previously[95]. Microcrystalline LiNc-BuO particles were suspended in 10mg/ml of MEM. A sonicator was used to create particles with sizes <200nm by pulsing the sample 10 times for 30 seconds each, with 1 minute cooling time (on ice) between successive pulses. The sample was placed on ice for 10-15 minutes to allow heavier, larger particles to settle. The supernatant liquid was then aliquoted out. For treatment with the EPR particles, both HET and SEG were grown to 70% confluence in 35mm dishes, then suspended in 1ml MEM. To this suspension, 50 μL of the LiNc-BuO suspension was added and the cells were incubated for 48 hours. The media was then replaced, the cells were resuspended and 20 μL of the suspension (~4000 cells) was drawn into a glass tube. The tube was placed in a flat cell resonator that was carefully aligned in the microwave cavity of an EPR spectrometer. EPR spectra was acquired for both cell lines as specified in the manufacturers guidelines. A previously reported calibration was used to convert spectra bandwidth to oxygen levels (0% $[O]_2$ = 210 mG line width, 8.5 mG/mmHg oxygen sensitivity for LiNc-Buo) [95].

4.3 Results

4.3.1 RTDP-FLIM Analysis of Cellular Lysate

The analysis from previous results and lysate-FLIM studies are shown in Table 4. Both HET-1 and SEG-1 are exposed to similar oxygen levels (227μM in solution), yet exhibit different RTDP lifetimes. Given that both lysates were generated in an identical fashion, differences in lifetime are likely indicative of intracellular biochemistry and of greater dynamic quenching in SEG vs. HET. The difference is small compared to RTDP lifetime (~13.4/300ns, or approximately 4.5%) and is statistically significant ($p<0.05$).

Sample	$\Delta\tau$ (ns)	$[O_2](\mu M)$	revised $\Delta\tau$ (ns)	$[O_2]_{Fl}$ (μM)	$[O_2]_{EPR}$ (μM)	Estimated K_q
HET cells		182.1		182.1	180.0	0.0044
SEG cells	44(a)	260.5	30.6	225.8	210.1	0.0054

HET-lysate		227
SEG-lysate	13.4(b)	227

Table 4: Lifetime differences and oxygen data from FLIM experiments on both living cells as well as for cellular lysates (data in grey boxes). All lifetime differences were computed as $\Delta\tau = \tau_{HET} - \tau_{SEG}$. Revised values of $\Delta\tau$ were computed as (a-b) and used to correct $[O_2]_{SEG}$ levels. Quenching Constant K_q values were re-estimated via the Stern Volmer equation with known lifetime and corrected oxygen values for SEGs.

FLIM in living cells indicates an average difference of 44ns between the intracellular RTDP fluorescence in HET and SEG cell lines, which lead to estimates of $[O_2]_{HET} = 182\mu M$ and $[O_2]_{SEG} = 260\mu M$. Given that extracellular oxygen is approximately $227\mu M$ and no known mechanisms of active oxygen transport exists in cells, it is likely that $[O_2]_{SEG}$ is overestimated by the current calibration. Indeed, lysate studies indicate that the potential lifetime (and corresponding) difference might have been overestimated by 13.4 ns. When corrected by this factor, we obtain $\Delta\tau = 30.6$ ns and a revised value of $[O_2]_{SEG} = 225 \ \mu M$. When compared to extracellular oxygen levels of 227 μM, this value is logical.

4.3.2 EPR Oximetry

EPR oximetry provides an established means of validating intracellular oxygen studies. Lysate studies on the experimental probe LiNc-BuO indicate little or no effect of the intracellular environment [95]. As indicated in Table 4, EPR oximetry studies on living HET and SEG cells yields several significant results. The HET oxygen levels compare favorably with fluorescence results, while the corrected SEG oxygen levels are similar to those reported by EPR. It is worth noting, however, that the trend of $[O_2]_{SEG} > [O_2]_{HET}$ is conserved, despite some expected variation due to the differences between the fluorescence imaging and EPR spectroscopic methods (e.g. adherent vs. suspended cell sample). Fig. 14 illustrates the two methods of oxygen sensing (FLIM and EPR), and the data yielded by each.

4.4 Conclusion

The application of spin-label oximetry to biological systems dates back almost 35 years, and EPR oximetry has come a long way since. Several other techniques exist for oxygen sensing, including (but not limited to) polarographic oxygen electrodes, PEBBLEs and EF5 binding. With an emphasis on measuring intracellular oxygen, EPR and fluorescence remain the only techniques currently capable of this. EPR, with the use of established and patient-friendly oximetry probes such as India Ink, remains the tool of choice. It is worthy to note that all the touted benefits of EPR are applicable to FLIM studies of oxygen as well; intensity independence (via peak-to-peak measurements for EPR and lifetime for FLIM) and true intracellular capability. In fact, while FLIM is applicable for both imaging and spectroscopic measurements of oxygen, EPR remains a spectroscopic tool. EPR imaging systems have been reported, but these are custom instruments with low resolution and (to our knowledge) not available commercially [109-111].

Cellular lysate has long been demonstrated to retain 'molecular viability' and have hence been used for functional studies of enzymatic activity as well as for protein extraction. As such, they form a logical starting point when ascertaining intracellular biochemical differences between two different biological species. Given our results, it is now possible to define a new protocol for intracellular oxygen sensing in living cells by using a 'reference cell line' with known oxygenation. For this work, that purpose is served by the

HET cells due to their stable oxygen levels across both fluorescence and EPR studies. However, any established cell line can be used in this role. For example, Chinese Hamster Ovary (CHO) cells, commercially available from ATCC have been used for EPR oxygenation studies with reproducible results ($[O_2]_{CHO}$ = 180 μM) [94].

A sample protocol, extensible for any oxygen-sensitive fluorophore, would be as follows:

1. Measure τ_0 for the fluorophore. Incubate the reference cells (with known oxygenation $[O_2]_{ref}$) with the fluorophore and measure τ_{ref}.
2. Calculate K_{ref} using the Stern Volmer equation for collisional quenching of fluorescence by oxygen, $\tau_0/\tau_{ref} = 1 + K_{ref}[O_2]_{ref}$.
3. Measure τ_{exp} for the experimental cell line with unknown oxygenation $[O_2]_{exp}$.
4. Correct $\tau_{exp} \rightarrow \tau_{exp\text{-}corr}$ with lysate studies using both the reference and experimental cell line.
5. Use the Stern Volmer equation to evaluate oxygenation $[O_2]_{exp}$ as $\tau_0/\tau_{exp\text{-}corr} = 1 + K_{ref}[O_2]_{exp}$.
6. Further, it is also possible once $[O_2]_{exp}$ is known to calculate a corrected $K_{exp\text{-}corr}$ that is specific to the experimental cell line. This can be calculated as $\tau_0/\tau_{exp} = 1 + K_{exp\text{-}corr}[O_2]_{exp}$. Since $K_{exp\text{-}corr}$ now contains the cell-specific biochemical information, there is no need to correct τ values for future experiments.

As compared to previous calibration efforts, this approach is a bit more time consuming, but beneficial for using a living reference biological sample, as well as accounting for intracellular biochemical information. Fluorescence imaging of oxygen provides a quantitative tool for metabolic studies using routine laboratory instrumentation that complements high resolution fluorescent detection of other metabolic indicators such as NADH, Ca^{2+}, ATP, etc. A fluorescence lifetime approach, while requiring advanced electronics, ensures that intensity based artifacts are minimized.

Chapter 5   OXYGEN MONITORING FOR CONTINUOUS CELL CULTURE

5.1 Introduction

Microfluidic devices have promising applications in cell-based assays and microscale tissue engineering, where spatio-temporal conditions are readily manipulated. Recently, PDMS-based microfluidic systems have been developed as biocompatible and rapidly prototyped systems for microscale-cell culture. For example, cells could be seeded and cultured successfully under continually perfused conditions to achieve an extracellular fluid-to-cell (volume) ratio close to the physiological value of 0.5 [112]. This small ratio facilitates heterogeneous chemical distribution, which may be critical in specifying cell fate in developing tissues. It is hence of great interest to quantitatively and with minimal perturbation characterize components (e.g., mitogens, nutrients, oxygen) in microfluidic bioreactors that influence cellular responses.

By association, it is necessary to develop tools to enable quantitative real-time control of microenvironment of cell culture in microbioreactors. This approach requires three main components: (i) the ability to actuate the spatio-temporal distribution of nutrients, growth factors, and adhesive signals in the cellular microenvironments, (ii) the ability to sense/measure nutrients, metabolites, growth factors, cytokines, and other cell-secreted products, and (iii) the ability to quantitatively model the relationship between various design and operating parameters to enable control and operation of the bioreactor. Our previous studies with development of highly versatile computerized microfluidic bioreactor arrays (Gu et al., 2004) are hence still incomplete in that they lack integration of chemical sensors and quantitative understanding of the bioreactors involved through modeling. This paper, alongwith other published work specifically addresses these issues for a crucial cell substrate, oxygen, by developing an optics-based method to quantify dissolved oxygen in PDMS microbioreactors and by applying a quantitative mathematical model to explain how key parameters control the spatial distribution of oxygen inside them [113-115].

Oxygen in cell cultures influences cell signaling, growth, differentiation, and death [112]. PDMS bioreactors are popular due to their high diffusivity of oxygen, which has been repeatedly demonstrated [116]. It has been observed, however, that diffusivity of PDMS can vary, depending on protein adsorption (e.g. when cells are cultured) or surface modification (e.g. plasma oxidization for bioreactors) [117]. It is hypothesized that this variability in PDMS permeability, along with cellular uptake and culture media perfusion, can affect spatial variations in oxygen within PDMS bioreactors.

Optical measurements of oxygen sensitive agents have advantages over more traditional, electrode-based approaches that make them uniquely applicable for bioreactor systems: they are well suited for small volumes, are relatively non-perturbing, and do not consume oxygen during the measurement. For long term cell culture, optical sensing enables time-lapse studies (hours or days) without disturbing the set-up, as well as imaging spatial oxygen distributions, which is useful for long-term cell culture.

Fluorescence intensity studies based on the oxygen-sensitivity of ruthenium complexes embedded in matrix have been performed [118, 119]; these studies were intensity-based, did not employ imaging, and did not report local oxygen concentration variations. Intensity-based fluorescence measurements are sensitive to instrumental variations (changes in excitation intensity or optical loss) and are affected by fluorophore concentration and photobleaching. Fluorescence lifetime, however, is an intrinsic property of the fluorophore's excited electronic state and is insensitive to intensity artifacts. FLIM in microfluidic structures was reported for studying solvent mixing by monitoring viscosity in channels without cells [120]. In this work, calibration of the oxygen sensitivity of RTDP lifetime [36] on a unique wide-field, time-domain FLIM system [59, 70] was applied for quantitative oxygen estimation. To our knowledge, this is the first demonstration of using a lifetime imaging modality for extracellular oxygen monitoring in PDMS bioreactors containing living cells.

5.2 Materials and Methods

5.2.1 FLIM and Quantitative Oxygen Sensing

The FLIM system and RTDP calibration was described earlier. RTDP calibration at 22°C [90] was verified using FLIM and an oxygen sensor (FOXY, Ocean Optics) [36]. RTDP fluorescence quenching by oxygen is a collisional process described by the Stern-Volmer equation: $\tau_0/\tau_x = 1 + K_q[O_2]_x$, where τ_0 = uninhibited RTDP lifetime (i.e. 0% oxygen), τ_x = RTDP lifetime at oxygen level $[O_2]_x$, and K_q is the Stern-Volmer quenching constant. For our measurements, K_q was determined to be = 2.7×10^{-3} μM^{-1} [36]. RTDP fluorescence images from bioreactor loops were collected by a 10X, 0.3NA objective (Zeiss, Jena, Germany) at 600 nm (HQ500lp, Chroma Technology Corp.) emission via 460 nm excitation at gates of 40, 140, 240, 340, and 440 ns, with the intensifier gate width set to 100 ns. A lifetime macro applying the rapid lifetime determination (RLD) algorithm was used to generate lifetime images on a per pixel basis from gated intensity images. Oxygen levels were ascertained by fitting the Stern-Volmer equation to calculated lifetimes at each pixel to generate oxygen distribution images (Fig. 16).

5.2.2 Bioreactor Fabrication and Cell Seeding

Each PDMS bioreactor was fabricated using backlight soft photolithography [121]. Each chip is composed of poly(dimethylsiloxane) (PDMS) and fabricated by using soft lithography (29). Prepolymer (Sylgard 184, Dow-Corning) at a 1:10 curing agent-to-base ratio was cast against positive relief features to form a flat, 1-mm-thick, negative replica for channel features. The relief features were composed of SU-8 (MicroChem, Newton, MA) and fabricated on a thin glass wafer (200 μm thick) by using backside diffused-light photolithography (30). The prepolymer was then cured at 60°C for 60 min, and holes were punched in it by a sharpened 14-gauge blunt needle. Channel indentations in the negative replica were sealed by a flat PDMS sheet. This sheet was 140 μm thick and formed by spin coating (200 rpm, 4 min) of a 1:10 ratio prepolymer onto glass wafers that were silanized with tridecafluoro-1,1,2,2-tetrahydrooctyl-1-trichlorosilane (United Chemical Technologies, Bristol, PA). This sheet was cured at 150°C for 60 min before sealing channels. An additional reservoir level was added only for cell culture devices. This layer was formed by adding prepolymer on machined brass molds to depths of 1 cm. This component was then cured at 60°C for 60 min. To irreversibly seal all cured components, they were oxidized for 30 s in oxygen plasma and sealed together.

The seed channel was injected with fibronectin to promote cell attachment. C2C12 cells (mouse myoblasts) were cultured under standard conditions and injected into the channels after being suspended in PBS [112]. RTDP (3 mg/ml) dissolved in PBS was injected into the media reservoir 1-2 hours after cell seeding and the bioreactor was perfused by peristaltic pumping action of an array of pin actuators adapted from Braille displays for three more hours to allow RTDP to distribute along bioreactor channels [112].

5.2.3 Bioreactor Imaging and Computational Validation

For imaging, the bioreactor was removed from the pumping system and placed on the microscope stage. RTDP yielded a bright signal within the PDMS bioreactor, which by itself was optically transparent. RTDP was well tolerated by the myoblasts within the bioreactor, as evidenced by cytotoxic and phototoxic studies done using Trypan Blue (Sigma-Aldrich). A published partial differential equation (PDE) model describing oxygen diffusion, convection, and uptake by cells within the device was developed and solved using FEMLAB (Consol AB Inc.) [114].

5.3 Results

5.3.1 Effect of Cell Density

The cell seeding procedure resulted in variable cellular densities across different microfluidic devices, so a basic bioreactor design was used to study effects of cell density on oxygen levels.

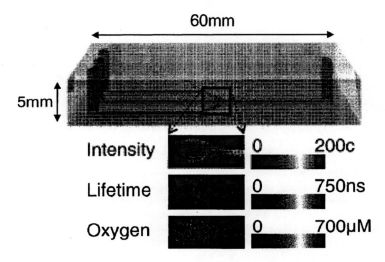

Figure 16. Illustration of fluorescence intensity and lifetime imaging in microfluidic devices. Top: perspective view of a device that contained C2C12 mouse myoblasts and was perfused with media containing RTDP at a rate of approximately 0.5 nl/s by gravity-driven flows. Channel height = 50 μm, width = 300 μm. Bottom: representative images of RTDP fluorescence intensity (scale in counts), lifetime (microseconds), and oxygen (μM) obtained via FLIM.

Fig. 16 illustrates the 3D design as well as typical intensity, lifetime, and oxygen images that were obtained via FLIM. The results from FLIM studies and computational simulations (accounting for channel geometry) were in good agreement and are shown in Fig. 17. As expected, the observed extracellular oxygen level in the bioreactor decreased in a cell-density dependent manner by almost 35% after incubation for 2 hours.

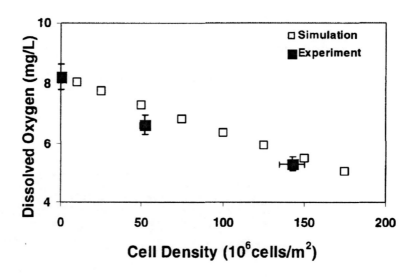

Figure 17. Simulation (white squares) and experimental FLIM (red squares) results of oxygen levels vs. cell densities in channels illustrated in Fig. 16. Oxygen levels were estimated by averaging pixel values in oxygen distribution images of the channel. The model simulations were carried out according to the equations described in [114], with the model parameters set at: maximum oxygen uptake rate $V_{max} = 2e^{-16}$ mol/cell/s; oxygen level at half-saturation $K_m = 0.0059$ mol/m^3; overall mass transfer coefficient $k_{la} = 4.5e^{-7}$ m/s, and estimated velocity of gravity flow $<u> = 0.003$ m/s. Error bars for some experimental data were within the red squares.

Oxygen levels measured here are lower compared to conventional macroscopic cell culture systems due to higher cell densities and low media levels. Lowered oxygen levels are still conducive to cell growth *in vitro*, as indicated by proliferation of cells in the bioreactor over a 12-14 hour period (data not shown).

5.3.2 Heterogeneity of Oxygen Distribution

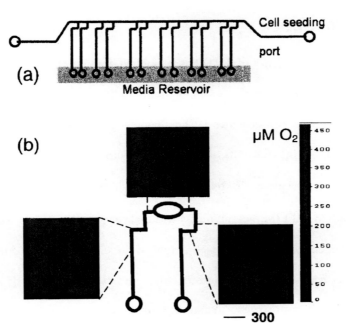

Figure 18. FLIM-based oxygen measurements from a closed-loop PDMS bioreactor for continuous cell culture of C2C12 mouse myoblasts. a) Device schematic. Channel shape was an isosceles trapezoid with a height of 30 μm and an upper (lower) PDMS layer of 180 μm (402 μm). Each of the six loops has a right and left valve separating it from the others. b) Oxygen distribution images at different points of a single loop (binary scale in μM).

A PDMS cell culture device with six loops is illustrated in Fig. 18a. The device was small scale, optically transparent and ideal for microscopic analysis. FLIM images of oxygen levels were taken at different points along each loop as illustrated in Fig. 18b. FLIM detection revealed that oxygen levels differed by as much as 20% within a loop and this trend was consistently observed across multiple devices. Differences in measurements made at different points within the same loop were statistically significant (ANOVA and student's t-test, $p < 0.001$). Comparisons across loops are not logical since each loop can potentially harbor different cell densities, which, as shown earlier, can significantly affect oxygen levels.

5.4 Conclusion

In conclusion, FLIM of RTDP was applied for the first time to measure oxygen in microfluidic bioreactors containing living cells. Measured extracellular oxygen levels correlated with cell densities in the channels and were verified with model simulations (Figs. 16,17). Initial results on a bioreactor with six loops (Fig. 18) revealed statistically significant variations in oxygen levels within each bioreactor loop. An implicit assumption in FLIM measurements was that oxygenation did not vary significantly in the axial dimension due to the small height of the channel cross-section (30 µm), an aspect that has been explored in more detail in simulation scenarios [114]. Future work will involve long-term oxygen monitoring under continually perfused conditions. Modeling results indicate that media recycling might be beneficial to retain growth factor(s) secreted by cells [114]; such a scheme will potentially affect oxygen levels, which can be tracked almost real-time due to fast data acquisition (<10 s) by the FLIM approach. All these endeavors are ultimately aimed at controlling long-term cellular responses (such as differentiation) in PDMS micro-bioreactors. A molecular, fluorescence lifetime-based approach to oxygen sensing in PDMS bioreactors, such as that described here, offers several advantages including replacing a macroscopic chemical electrode with a non-destructive (and minimally perturbing) optical imager, which could potentially be customized and scaled-down using optoelectronic technologies, making this approach more economical and portable.

These results are important not only for characterization of the specific device we describe in this report but also for formulating general guidelines for designing PDMS devices in regards to their oxygen microenvironment for mammalian cell culture. In related work, the relationship that describes the minimum flow rate that will offset specific oxygen consumption rates will be useful to the microfluidics community for designing devices and experimental conditions for microfluidic cell culture. Our results also underscore the importance of characterizing the oxygen transfer rate of each type of PDMS device made under different conditions such as plasma oxidation and protein coating to explain cell behaviors observed as the oxygen permeability of a given device cannot always be predicted by simply using the permeability values of native bulk PDMS. The methods described here also provide an efficient method to perform such analyses.

Some drawbacks exist to this approach. Significantly, high levels of RTDP are required to achieve respectable SNR, which could affect cellular viability. This problem can be circumvented by more sensitive detection, or using a higher NA objective, both of which would allow for lower RTDP levels to be used. It is also quite possible that the RTDP signal recorded is a hybrid intra/extracellular RTDP signal, which could lead to misleading lifetime and oxygen values. We do not anticipate this to be the case with our experiments due to the relatively quick time frame, but it should be accounted for when more long-term studies are performed. Lastly, no temperature control was present for the studies performed; a temperature controller for the bioreactor has now been developed, and should provide more realistic oxygen level estimates for future experiments.

Advances in FLIM and spatial resolution could supplement our results with high-resolution (~single cell) imaging of intracellular oxygen. As described before, individual cells can be incubated with RTDP, which can be washed away from the media to provide exclusively intracellular RTDP fluorescence signal. Without the high background signal from media, a high resolution objective (100x for example) could be used for oxygen estimation in single cells in a 3D arrangement, similar to tissues. 3D imaging with wide-field systems, however, raises the issue of resolution degradation due to out-of-focus fluorescence, an issue we try to address in the next chapter.

Chapter 6 IMAGE RESTORATION IN FLIM

6.1 Introduction

Fluorescence lifetime is defined as the rate constant associated with an exponentially decaying fluorophore population [30]. Given the independence of fluorescence lifetimes from intensity associated artifacts such as concentration, scattering and absorption, FLIM is uniquely applicable for high resolution studies of biological systems and their milieu. Lifetime studies have hence been used to study physiological parameters such as (but not limited to) membrane dynamics, metabolism, oxygenation and molecular associations [36, 81, 122, 123].

While theoretically independent of intensity, several fluorescence lifetime imaging microscopy (FLIM) techniques derive lifetime images via intensity image analysis [124]. As a result, quality of the lifetime map is directly affected by the acquisition of intensity images. It is commonly known that the imaging properties of any optical microscope give rise to distortions [125].

FLIM is increasingly applicable for studies of intrinsic fluorescence, with an eye on clinical applications. Naturally-occurring fluorophores such as NADH, collagen, keratin, flavins, porphyrins, etc. provide promise for non-invasive analysis of living cells and tissues. A key issue is that the low quantum yield of such fluorophores, coupled with the use of low energy and pulsed laser sources, results in weak, barely detectable signals [126]. This necessitates the use of signal amplification techniques, such as Intensified-CCD cameras, that are a key component of several wide-field FLIM systems [70, 127, 128]. The resulting images, however, suffer from haze and corresponding loss of spatial resolution. Fig. 19 illustrates the effect of an intensifier on a high SNR biological image with almost no background, yet significant haziness or 'lateral smearing' is observed. The occurrence of haze due to image intensifiers is explained by several reasons, including low gating voltages for ultrafast intensifiers and degradation by the phosphor screen [129].

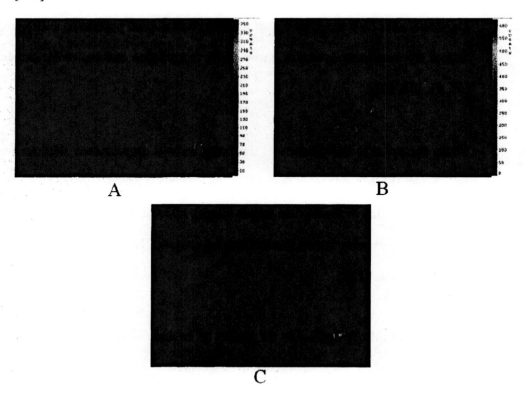

Figure 19: Illustration of lateral smearing, or haze, with an image-intensified CCD camera. A) Blue fluorescence from a fixed mouse intestine section imaged with a CCD alone. B) Same region when imaged with an ICCD. The demagnification due to the lens-coupling between the intensifier and CCD (=2.17) is evident in the image. C) Red rectangular region from (B) magnified to show loss of resolution. The excitation source for all images was a mercury lamp.

The operation of an ICCD camera is shown in Fig. 20. Signal received from the optical system (microscope) is converted to photoelectrons by the photocathode. Note that ultrafast gating of the ICCD is possible by applying a bias voltage directly to the photocathode; a negative voltage will inhibit ICCD operation since the photoelectrons will not enter the microchannel plate (MCP). When a positive bias is applied, the electrons fly to the MCP, which is equivalent to an array of PMTs and consists of hollow channels lined with the appropriate material for signal amplification. The extent of amplification is controlled by the MCP voltage. The photoelectrons are then converted back to photons by the phosphor screen. Finally, the output of the phosphor screen is lens-coupled to the CCD.

Intensifier distortion occurs primarily at the photocathode; photoelectrons released here take divergent paths, and a high voltage must be applied to make them enter the appropriate MCP channel. However, it is very difficult to switch high voltages at ultrafast speed (>100 MHz), and rapid overheating can occur. Hence, lower voltages are used at the cost of resolution, which occurs due to photoelectrons spilling into neighboring MCP channels, resulting in lateral 'smearing' of the image.

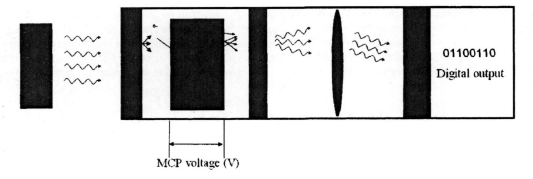

Figure 20: Illustration of ICCD operation. The CCD camera is denoted by the CCD chip at the end of the diagram which provides the digital readout to the PC; all other components are part of the image intensifier.

While haze can be qualified as unwanted signal or noise, it is inherently more difficult to remove while maintaining the quantitative relationships within the intensity image (for accurate lifetime estimation). Further, since the source of haze is the signal itself, it has a near-identical lifetime and is virtually indistinguishable in lifetime images. This leads to lateral smearing in lifetime maps as well, where the footprint of the fluorescent structure is larger in the lifetime image as compared to intensity images, leading to a loss of resolution.

Limited work has been performed towards computational image restoration in FLIM. Among optical methods, structured illumination has been used to improve spatial resolution of intensity and lifetime maps based on the Moiré effect, albeit at the loss of SNR, which is an issue when working with dim fluorophores [130]. Squire et al. illustrated the possibility of reconstructing 3D lifetime images from a frequency-domain FLIM system via the established Iterative Constrained Tikhonov-Miller (ICTM) algorithm [76]. Intensity-based image restoration, on the other hand, has seen extensive interest, both research and commercial. Approaches include prior knowledge of the object, known or estimated noise characteristics, known or estimated PSF, constrained methods, etc [131-135].

In this work, an iterative constrained 2D blind deconvolution algorithm was applied to FLIM intensity images (direct lifetime restoration), and the effect on lifetime images and values were observed. By combining both restored-intensity and lifetime maps, a new approach to fluorescence lifetime representation, intensity-overlay restoration, was proposed. Utilization of current intensity-based image restoration techniques to FLIM provides promise for more lifetime-specific restoration algorithms in the future. With advances in desktop computing power, numerical methods provide increasingly better resolution without compromising SNR (sometimes improving it) or the need for additional optical hardware, and are readily applicable to previously acquired data as well.

For experimental purposes, a wide-field, time-domain fluorescence lifetime imaging microcopy (FLIM) system was utilized to probe fluorescence in microspheres, living cells and fixed tissues.

6.2 Materials and Methods

6.2.1 Sample Preparation

Fluorescent Microspheres – Yellow-green fluorescent microspheres were purchased in three sizes of 1µm, 3µm, 10µm diameter (Polysciences, Warrington, PA) and each was diluted in PBS to a concentration of approximately 10^4 spheres/ml. 20µl of each was dried between a glass slide and coverslip for imaging. All YG spheres has an excitation/emission maximum of 441/486nm as specified by the manufacturer.

RTDP stained cancer cells – RTDP (460/600nm) incubation and temperature-controlled imaging of living Barrett's adenocarcinoma columnar cells (SEGs) has been described earlier [36].

Fixed Mouse Intestine section – A fixed cryostat mouse intestine section of approximately 16 µm thickness was purchased (Molecular Probes, Carlsbad, CA). The slide is multiply-stained; Alexa Fluor 350 wheat germ agglutinin (346/442nm), a blue-fluorescent lectin, was used to stain the mucus of goblet cells. The filamentous actin prevalent in the brush border was stained with red-fluorescent Alexa Fluor 568 phalloidin (578/600). The nuclei were stained with SYTOX Green nucleic acid stain (504/523).

6.2.2 Image Acquisition and Analysis

4 gated images (i.e. taken at different delays) were acquired via FLIM and analyzed to yield lifetime maps as described earlier [70]. Adaptive noise removal, as described earlier, was used for all image analysis.

6.2.3 Image Restoration

Intensity image restoration of each gated image was carried out by the Autoquant (Media Cybernetics, Bethesda, MD) software. For the purpose of this study, the adaptive PSF estimation capability of the algorithm was used. The underlying Maximum Likelihood Estimation (MLE) algorithm, a mathematical optimization strategy that is generally used for producing estimates of quantities corrupted by some form of random noise, forms the basis of the 2D Blind approach and has been explained elsewhere [136, 137].

For each images series (i.e. 4 gated intensity images), the first gated image was analyzed with the blind/derived PSF setting. The output was the restored image, as well as a PSF file. The PSF file was saved and used to analyze the next 3 gated images (i.e., non-blind). Each image was restored with 10 iterations and medium-high noise specification. For each image the optical parameters specified are listed in Table 5. Note that the pixel spacing parameter accounted for demagnification due to the image intensifier:

Parameter	Setting
Modality	Wide-field fluorescence
Pixel Spacing (in μm)	0.219 x 0.219 (100x)
	0.54 x 0.54 (40x)
	2.19 x 2.19 (10x)
Objective Magnification (NA)	100x (1.3)
	40x (1.3)
	10x (0.25)
Immersion Medium	Glycerin(RI = 1.47) for the 100x/40x
	Air (RI = 1) for the 10x
Emission Wavelength	sample-dependent

Table 5. List parameters used as input for computational image restoration.

The computationally-restored images were then analyzed for lifetime. Unless specified otherwise, all lifetime analysis was conducted using the adaptive haze removal technique.

6.3 Results

6.3.1 Computational Restoration of Gated Images

Development of computational methods that preserve quantitative aspects of images hold promise for lifetime studies. The 2D Blind algorithm is constrained to preserve total photon counts during computational restoration (personal communication). This is especially useful for analyzing ICCD-generated images, where the haze, instead of being removed, is reassigned to the (apparent) source of origin. Hence, not only is the haze in the image greatly reduced, but the SNR of the fluorescent source is increased.

Fig. 21A-D illustrates the effect of restoration on both intensity and lifetime images on 3 micron-diameter YG fluorescent microspheres. Overlap of haze from the 2 spheres (Fig. 21A) leads to a lifetime map where one is indistinguishable from the other (Fig. 20B). As expected, the restored intensity map (Fig. 21C) has improved SNR, spheres are brighter. The reduction in haze is evident in the lifetime map (Fig. 21D), where the 2 spheres are now clearly distinguishable. Important as well, there is only a 3% change in lifetime (from 1.65 to 1.60ns). It is also worthy to note that the edge effect that arises with adaptive noise removal is almost eliminated in, and might even explain the change in lifetime value.

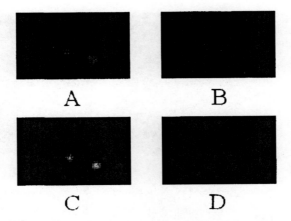

Figure 21. A,B: Native fluorescence intensity, lifetime images of 3-micron diameter YG spheres. C,D: Corresponding restored images. The previously indistinguishable pair of spheres are evident in the restored lifetime image, as is a reduction in edge pixels with large lifetime values.

Fig. 22A-D illustrates the effect of image restoration on a living biological sample. Again, image restoration leads to improved resolution (Fig. 22C), smaller footprint for the fluorescent structures (Fig. 22D), and similar lifetimes. The edge effect is also reduced, though the result is not as profound as for the microspheres. Applying a previously calculated RTDP lifetime calibration for oxygen sensitivity to this image will not affect the calculated oxygen levels, hence preserving the quantitative nature of the image. One artifact, however, that arises with the cell images is the presence of spindly features at the cellular boundaries. Another synthetic feature that manifests is the patchiness of the cells themselves. The patchiness was seen to arise with almost every biological sample studied (data not shown) and was not evident in any of the intensity images, native or restored. This indicates it is likely a product of the RLD analysis rather than an actual, underlying biological structure.

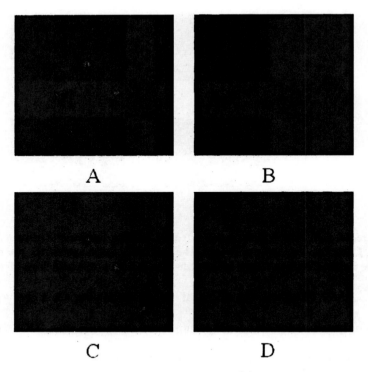

Figure 22. A,B: Native fluorescence intensity, lifetime images of RTDP-incubated SEGs. C,D: Corresponding restored images.

6.3.2 Weighted Intensity-Lifetime Mapping

Both fluorophore intensity and lifetime provide complementary information; while lifetime is indicative of the milieu of the fluorescent molecule, intensity provides concentration, and sometimes morphological information. It has previously been suggested that whether a given pixel contributes to the information in an image should be weighted by both its lifetime and intensity content [138]. Such an approach attenuates haze to a great extent due to its low SNR, even though haze has virtually the same lifetime as the source. It is now possible to use advances in image restoration to significantly improve the quality of such weighted intensity-lifetime maps. By simply restoring the first gated image (highest SNR) alone and using it to weigh the native lifetime image, we can obtain a highly resolved lifetime distribution image for the fluorescent beads. A key strength of some of the robust intensity-based restoration algorithms (such as 2D Blind) is they seldom create structural artifacts within a given image. By using the native lifetime image (as opposed to one derived from restored intensity images), the lifetime/quantitative content has been fully conserved as well. In fact, due to de-emphasis of the edge pixels, the spiky lifetime values (red pixels) are missing from the restored I-τ maps, thereby yielding a more realistic lifetime distribution. One approach to do this is illustrated in Fig. 23 for a star arrangement of 10-micron YG fluorescent beads. The intensity images are represented in gray scale (Fig. 23A,C), while the lifetime image is color coded (Fig. 23B). The haze from the five spheres is significant, resulting in a single lifetime blob. A simple product of the two images, though, yields a stark improvement in resolution (Fig. 23D). Every pixel in the resulting image is defined by brightness=intensity and color=lifetime.

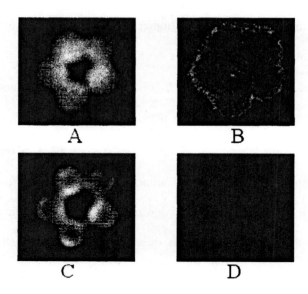

Figure 23. A: Native fluorescence intensity of five 10-micron YG beads. B. Native fluorescence lifetime map C: Restored intensity image. Note the reduction in haze. D. Restored intensity weighted-lifetime map.

Fig. 24 again illustrates how the I-τ approach can be applied to study only relevant content in biological systems without sacrificing SNR. The mucus of goblet cells are stained with Alexa Fluor 350, a blue-emitting fluorophore that is excited by UV light. Since most endogenous fluorophores also excite/emit in the UV/blue, a dim background fluorescence from the tissue is also observed. While it is barely detectable in the intensity image (Fig. 24A), it is clearly evident in the lifetime map (Fig. 24B) and has comparable lifetime to the stained cells, making them difficult to visualize. The hybrid map, however, restored the focus on the lifetime of the brightly fluorescent cells (Fig. 24D). Using the restored intensity images provide even better resolution without generating any structural anomalies.

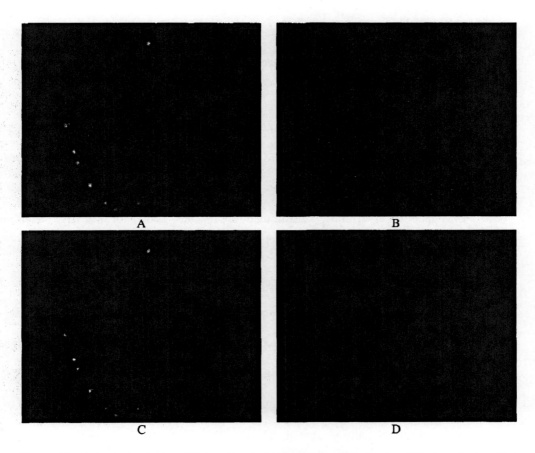

Figure 24. A: Low resolution (10x) native intensity image of a fixed mouse intestine section exhibiting Alexa 350 fluorescence. B. Native fluorescence lifetime map. C. Restored intensity image. D. Restored intensity weighted-lifetime map.

6.4 Conclusion

Wide-field microscopes are routinely used to collect 2D data sets. Such images contain sharp in-focus features as well as blur due to signal originating above or below the plane of focus. With the use of ultrafast gated image intensifiers for FLIM, further in-plane smearing (haze) is observed, which leads to reduced resolution for intensity and especially for lifetime imaging. To recover true image heterogeneity, it is advantageous to not only eliminate some of the haze, but to be able to reassign it to its source, so as to improve SNR. This article reports results from an approach involving resolution enhancement via computational image restoration in wide-field FLIM.

Direct image restoration of gated intensity images is the most straightforward approach for improving resolution of FLIM images. Restoration techniques have been in development for >20 years now and form a mature foundation to base future work on FLIM-specific algorithms. Constrained approaches allow for selective retention of relevant quantitative information in a fluorescence image while improving resolution. Further, blind deconvolution provides the convenience of not requiring the PSF of the optical system, but generating a reconstructed estimate instead, also with constraints [133]. PSF measurement via fluorescent beads, as is the most commonly used approach, suffers from potential photobleaching effects and statistical noise due to low SNR [139]. The PSF is also distorted by variations in refractive index in biological samples, which makes computationally-generated, adaptive PSF approaches more relevant. This work indicates that iterative, constrained 2D blind deconvolution holds potential for improving spatial resolution in wide-field FLIM by using restored intensity images for evaluation fluorescence lifetimes via the RLD approach. The commercial availability of this algorithm makes it easy to adopt. Image restoration also lead to reduction in noise, an observation explained by the inherent assumption of quantum photon noise in MLE. Adaptive haze removal further improves image quality. Future work in this regard could involve development of FLIM specific constraints that correct some of the observed structural artifacts associated with biological samples.

Correlation of fluorescence intensity-weighted lifetime images has been reported previously as a means to provide morphological (intensity) and functional (lifetime) information in the same image, and also improves FLIM spatial resolution. A significant enhancement, however, is afforded by using restored intensity images to balance the native lifetime image. The benefit of this approach is that not only quantitative algorithms like the 2D blind, but any restoration/deblurring algorithm can be used without sacrificing lifetime information. Note that all the data presented in this work is largely 2D, as is the analysis. We also assume the veracity of all lifetime information obtained. In reality, there is always the prospect of lifetime 'blurring' due to out of focus haze. Removal of haze due to the 3D nature of thicker samples (and its effect on fluorescence lifetime) has been reported in a limited sense so far and provides impetus for future work in the 3D blind restoration-FLIM regime. Wide-field FLIM has its limitations for studying 3D samples, especially those with high background [140]. Restoration techniques hold promise for extending these limits without the use of additional optics and while increasing SNR.

Chapter 7 CONCLUSIONS AND FUTURE WORK

7.1 Conclusions

The primary goal of this dissertation was to develop resolution enhanced-fluorescence lifetime imaging microscopy (FLIM) as an effective tool for accurate studies of metabolic function via combined endogenous (NADH) and exogenous (oxygen) sensing. It was hypothesized that intracellular oxygen sensing was possible by not only calibrating the lifetime dependence of RTDP on oxygen, but also that verifying (via EPR studies) and correcting (via lysate-FLIM studies) this process would yield an accurate estimate of oxygenation. Resolution enhancement via computational means provides the possibility of extending wide-field FLIM capability deeper into the 3D realm.

In **Chapter 2**, we described the concept and instrumentation behind our wide-field time-domain FLIM system. Lifetime data analysis was carried out via the Rapid Lifetime Determination (RLD) approach and was extensible to an arbitrary number of gates. Noise/haze removal was achievable by two separate approaches, adaptive and baseline, each with its own benefits and drawbacks. The adaptive approach was utilized for all experiments thereafter.

In **Chapter 3** we developed a quantitative, calibrated approach to oxygen sensing using tris(2,2'-bipyridyl)dichloro-ruthenium(II) hexahydrate (RTDP). The calibration process was temperature controlled and carried out with a fluorescent optical sensor, which minimizes artifacts. Our approach is applicable to a wide range of oxygen estimation (0-300μM of oxygen) with high sensitivity (±8μM). We were able to show oxygen decrease in living cells enclosed in an airtight compartment. Intracellular oxygen in both living HET and SEG cells was measured using RTDP and FLIM. Mean oxygen levels were higher in the SEG vs. the HET.

We showed FLIM and confocal imaging of endogenous NAD(P)H in living normal (HET) and cancerous (SEG) esophageal cells. Intensity-based measurements indicated consistently higher NAD(P)H levels in cancer vs. normal cells, which is similar to clinical observations. Similarity of NAD(P)H lifetimes indicated the differences between the two cell lines was likely due to concentration differences. Mitochondrial stains were used to verify that the origin of endogenous fluorescence was the mitochondria and not the cytoplasm, indicating that NADH (and not NADPH) was likely the observable fluorescent entity.

Combined with previously obtained results for NADH fluorescence, it is likely indicative of mitochondrial dysfunction in SEG. Special FLIM properties hence allows for near-simultaneous sensing of nanosecond endogenous NAD(P)H (UV excitation and ultrafast gating) as well as exogenous oxygen sensing via long-lived RTDP fluorescence (single-shot operation).

Chapter 4 addressed a key concern with calibration for quantification of biological parameters; the validity of the calibration protocol, an accepted issue with oxygen sensing. Specifically, it is near impossible to replicate the intracellular environment as a media for calibration. We use a two-pronged approach (cellular lysate studies and EPR) to address this issue and correct our previously reported oxygen levels in HET and SEG. We use a solution of RTDP dissolved in cellular lysate of HET and SEG. We determined that RTDP lifetime differed between the two lysates when normalized for oxygen, indicating intracellular biochemical differences. The intracellular oxygen data was corrected based on lysate results. Then the revised oxygen levels were accurately verified via EPR studies, widely considered the gold standard for intracellular oxygen sensing. A generic protocol for correcting in vitro calibration, and the use of reference cell line, was proposed and is extensible to other oxygen-sensitive fluorescent probes as well. Given the significant differences in oxygenation observed after correction via lysate studies, it is evident that more sophisticated processed such as those proposed here are a necessary step for accurate quantification of intracellular biological parameters.

Chapter 5 - While RTDP-based FLIM finds great impact for intracellular oxygen sensing, it is almost as readily applicable for extracellular sensing as well. PDMS-based bioreactors, which are continuous cell culture microfluidic systems capable of culturing cells to near tissue-level densities, benefit enormously from better characterization of gradients (of oxygen, nutrients, mitogens, etc) that arise during such cultures. This is because in turn, oxygen levels can influence differentiation and viability of cultures cells.

RTDP dissolved in media and pumped through the device provides a simple application for FLIM. Oxygen levels were found to decrease with increasing cell density. This logical observation was verified with a computational model of oxygen diffusion generated using partial differential equations. The FLIM image was also observed to change at different points along flow loops within the bioreactors, with differences as high as 20% between end points. Since RTDP concentration was well-controlled, intensity-based studies were used to corroborate FLIM results and conduct further studies as well.

In **Chapter 6** we addressed the issue of noise and loss of resolution in ICCD-based FLIM systems, an area that has seen little improvement. Specifically, we addressed the problem of lateral smearing and haze induced due to image intensifier use. As a first step, we introduced the concept of user-defined baseline and adaptive haze removal, two approaches that are also beneficial for background removal. We exploited current advances in iterative constrained image restoration of fluorescence intensity images (2D blind deconvolution in particular) to further refine our approach by developing two different approaches; direct lifetime restoration, which entailed restoration of each gated image followed by lifetime map generation via the RLD approach described earlier, and intensity-overlay restoration, which was achieved by restoring the first gated image, and then using its intensity information to weigh the original lifetime map. Both approaches were shown to improve spatial resolution of FLIM maps in a variety of fluorescent samples, including microspheres, living cells, and fixed tissue samples, and indicate a significant advance for the field. An important concern was veracity of lifetime after the restoration process. FLIM restoration was shown to improve resolution without significantly affecting lifetime, and indicates the potential for further, FLIM-specific research.

The major contributions of this dissertation can be summarized as follows:

- Developed calibration and demonstrated capability of FLIM for quantitative oxygen sensing in living cells via RTDP.
- Developed a FLIM system with approaches for noise removal in FLIM images, baseline and adaptive.
- Demonstrated FLIM study of metabolic function in normal and cancer cells via a hybrid NADH (endo)-RTDP (exo) approach.
- Validation and further calibration of quantitative oxygen measurements by EPR and cell lysate studies. Proposed new protocol for accurate estimation of intracellular oxygen.
- Demonstrated FLIM study of oxygenation gradients in microfluidic bioreactors with 3D cell distribution.
- Developed new lifetime and intensity-overlay image restoration (blind) techniques for enhancing FLIM spatial resolution and extension of capability into the 3D realm.

The work in this dissertation has been presented and documented as cited below:

Chapter 2:

Light-Scattering Spectroscopy for Evaluating Dysplasia in patients With Barrett'e Esopghagus. *Gastroenterology* 2001, 120:1620-1629.

97. Glasgold R, Glasgold M, Savage H, Pinto J, Alfano R, Schantz S: Tissue Autofluorescence as an intermediate endpoint in NMBA-induced esophageal carcinogenesis. *Cancer Letters* 1994, 82:33-41.

98. Gibson BW: The human mitochondrial proteome: oxidative stress, protein modifications and oxidative phosphorylation. *Int J Biochem Cell Biol* 2005, 37:927-934.

99. Chen LB, Welss MJ, Davis S, Bleday RS, Wong JR, Song J, Terasaki M, Shepherd EL, Walker ES, Steele GD, Jr.: Mitochondria in living cells: effects of growth factors and tumor promoters, alterations in carcinoma cells, and targets for therapy. *Cancer Cells* 1985, 3:433-443.

100. Brogan AP, Widger WR, Kohn H: Bicyclomycin fluorescent probes: synthesis and biochemical, biophysical, and biological properties. *J Org Chem* 2003, 68:5575-5587.

101. Mamelak AJ, Kowalski J, Murphy K, Yadava N, Zahurak M, Kouba DJ, Howell BG, Tzu J, Cummins DL, Liegeois NJ, Berg K, Sauder DN: Downregulation of NDUFA1 and other oxidative phosphorylation-related genes is a consistent feature of basal cell carcinoma. *Exp Dermatol* 2005, 14:336-348.

102. Xu H, Aylott JW, Kopelman R: Fluorescent nano-PEBBLE sensors designed for intracellular glucose imaging. *Analyst* 2002, 127:1471-1477.

103. Simpkins C, Balderman S, Mensah E: Mitochondrial oxygen consumption is synergistically inhibited by metallothionein and calcium. *Journal of Surgical Research* 1998, 80:16-21.

104. Swartz HM: Using EPR to measure a critical but often unmeasured component of oxidative damage: Oxygen. *Antioxidants & Redox Signaling* 2004, 6:677-686.

105. Swartz HM, Khan N, Buckey J, Comi R, Gould L, Grinberg O, Hartford A, Hopf H, Hou HG, Hug E, Iwasaki A, Lesniewski P, Salikhov I, Walczak T: Clinical applications of EPR: overview and perspectives. *Nmr in Biomedicine* 2004, 17:335-351.

106. Evans SM, Judy KD, Dunphy I, Jenkins WT, Nelson PT, Collins R, Wileyto EP, Jenkins K, Hahn SM, Stevens CW, Judkins AR, Phillips P, Geoerger B, Koch CJ: Comparative measurements of hypoxia in human brain tumors using needle electrodes and EF5 binding. *Cancer Research* 2004, 64:1886-1892.

107. Gallez B, Swartz HM: In vivo EPR: when, how and why? *Nmr in Biomedicine* 2004, 17:223-225.

108. Grauw CJ, Gerritsen HC: Fluorescence lifetime imaging of oxygen in dental biofilm. *Proceedings of SPIE* 2003, 4164:70-78.

109. Ahmad R, Clymer B, Vikram DS, Deng YM, Hirata H, Zweier JL, Kuppusamy P: Enhanced resolution for EPR imaging by two-step deblurring. *Journal of Magnetic Resonance* 2007, 184:246-257.

110. Deng YM, Petryakov S, He GL, Kesselring E, Kuppusamy P, Zweier JL: Fast 3D spatial EPR imaging using spiral magnetic field gradient. *Journal of Magnetic Resonance* 2007, 185:283-290.

111. Matsumoto K, Subramanian S, Devasahayam N, Aravalluvan T, Murugesan R, Cook JA, Mitchell JB, Krishna MC: Electron paramagnetic resonance imaging of tumor hypoxia: Enhanced spatial and temporal resolution for in vivo pO(2) determination. *Magnetic Resonance in Medicine* 2006, 55:1157-1163.

112. Gu W, Zhu XY, Futai N, Cho BS, Takayama S: Computerized microfluidic cell culture using elastomeric channels and Braille displays. *Proceedings of the National Academy of Sciences of the United States of America* 2004, 101:15861-15866.

113. Mehta G, Mehta K, Sud D, Song J, Bersano-Begey T, Futai N, Mycek M-A, Linderman J, Takayama S: Quantitative Oxygen Measurements and Analysis of Cell Respiration in Microfluidic Bioreactors. *Submitted for publication to Lab on a Chip* 2006.

114. Mehta K, Linderman JJ: Model-based analysis and design of a microchannel reactor for tissue engineering. *Biotechnology and Bioengineering* 2006, 94:596-609.

115. Sud D, Mehta G, Mehta K, Linderman J, Takayama S, Mycek M-A: Optical Imaging in Microfluidic Bioreactors Enables Oxygen Monitoring for Continuous Cell Culture. *Journal of Biomedical Optics Letters* 2006, in press.

116. Leclerc E, Sakai Y, Fujii T: Cell culture in 3-dimensional microfluidic structure of PDMS (polydimethylsiloxane). *Biomedical Microdevices* 2003, 5:109-114.

117. Shiku H, Saito T, Wu CC, Yasukawa T, Yokoo M, Abe H, Matsue T, Yamada H: Oxygen permeability of surface-modified poly(dimethylsiloxane) characterized by scanning electrochemical microscopy. *Chemistry Letters* 2006, 35:234-235.

118. Gao FG, Jeevarajan AS, Anderson MM: Long-term continuous monitoring of dissolved oxygen in cell culture medium for perfused bioreactors using optical oxygen sensors. *Biotechnology and Bioengineering* 2004, 86:425-433.

119. Sin A, Chin KC, Jamil MF, Kostov Y, Rao G, Shuler ML: The design and fabrication of three-chamber microscale cell culture analog devices with integrated dissolved oxygen sensors. *Biotechnology Progress* 2004, 20:338-345.

120. Benninger RKP, Hofmann O, McGinty J, Requejo-Isidro J, Munro I, Neil MAA, deMello AJ, French PMW: Time-resolved fluorescence imaging of solvent interactions in microfluidic devices. *Optics Express* 2005, 13:6275-6285.

121. Futai N, Gu W, Takayama S: Rapid prototyping of microstructures with bell-shaped cross-sections and its application to deformation-based microfluidic valves. *Advanced Materials* 2004, 16:1320-+.

122. Konig I, Schwarz JP, Anderson KI: Fluorescence lifetime imaging: Association of cortical actin with a PIP3-rich membrane compartment. *Eur J Cell Biol* 2008.

123. Schweitzer D, Schenke S, Hammer M, Schweitzer F, Jentsch S, Birckner E, Becker W, Bergmann A: Towards metabolic mapping of the human retina. *Microsc Res Tech* 2007, 70:410-419.

124. Sharman KK, Periasamy A, Ashworth H, Demas JN, Snow NH: Error analysis of the rapid lifetime determination method for double-exponential decays and new windowing schemes. *Analytical Chemistry* 1999, 71:947-952.

125. Goodwin PC: Evaluating optical aberration using fluorescent microspheres: methods, analysis, and corrective actions. *Methods Cell Biol* 2007, 81:397-413.

126. Petty HR: Fluorescence microscopy: established and emerging methods, experimental strategies, and applications in immunology. *Microsc Res Tech* 2007, 70:687-709.

127. Cole MJ, Siegel J, Webb SE, Jones R, Dowling K, French PM, Lever MJ, Sucharov LO, Neil MA, Juskaitis R, Wilson T: Whole-field optically sectioned fluorescence lifetime imaging. *Opt Lett* 2000, 25:1361-1363.

128. Uchimura T, Kawanabe S, Maeda Y, Imasaka T: Fluorescence lifetime imaging microscope consisting of a compact picosecond dye laser and a gated charge-coupled device camera for applications to living cells. *Anal Sci* 2006, 22:1291-1295.

129. Pietka E, Huang HK: Correction of Aberration in Image-Intensifier Systems. *Computerized Medical Imaging and Graphics* 1992, 16:253-258.

130. Cole MJ, Siegel J, Webb SE, Jones R, Dowling K, Dayel MJ, Parsons-Karavassilis D, French PM, Lever MJ, Sucharov LO, Neil MA, Juskaitis R, Wilson T: Time-domain whole-field fluorescence lifetime imaging with optical sectioning. *J Microsc* 2001, 203:246-257.

131. Boutet de Monvel J, Le Calvez S, Ulfendahl M: Image restoration for confocal microscopy: improving the limits of deconvolution, with application to the visualization of the mammalian hearing organ. *Biophys J* 2001, 80:2455-2470.

132. Guan YQ, Cai YY, Zhang X, Lee YT, Opas M: Adaptive correction technique for 3D reconstruction of fluorescence microscopy images. *Microsc Res Tech* 2008, 71:146-157.

133. Pankajakshan P, Zhang B, Blanc-Feraud L, Kam Z, Olivo-Marin JC, Zerubia J: Parametric blind deconvolution for confocal laser scanning microscopy. *Conf Proc IEEE Eng Med Biol Soc* 2007, 2007:6532-6535.

134. Sibarita JB: Deconvolution microscopy. *Adv Biochem Eng Biotechnol* 2005, 95:201-243.

135. Vermolen BJ, Garini Y, Young IT: 3D restoration with multiple images acquired by a modified conventional microscope. *Microsc Res Tech* 2004, 64:113-125.

136. Markham J, Conchello JA: Fast maximum-likelihood image-restoration algorithms for three-dimensional fluorescence microscopy. *J Opt Soc Am A Opt Image Sci Vis* 2001, 18:1062-1071.

137. Verveer PJ, Gemkow MJ, Jovin TM: A comparison of image restoration approaches applied to three-dimensional confocal and wide-field fluorescence microscopy. *J Microsc* 1999, 193:50-61.

138. Siegel J, Elson DS, Webb SE, Lee KC, Vlandas A, Gambaruto GL, Leveque-Fort S, Lever MJ, Tadrous PJ, Stamp GW, Wallace AL, Sandison A, Watson TF, Alvarez F, French PM: Studying biological tissue with fluorescence lifetime imaging: microscopy, endoscopy, and complex decay profiles. *Appl Opt* 2003, 42:2995-3004.

139. de Monvel JB, Scarfone E, Le Calvez S, Ulfendahl M: Image-adaptive deconvolution for three-dimensional deep biological imaging. *Biophys J* 2003, 85:3991-4001.

140. Murray JM, Appleton PL, Swedlow JR, Waters JC: Evaluating performance in three-dimensional fluorescence microscopy. *J Microsc* 2007, 228:390-405.

141. Paxian M, Keller SA, Cross B, Huynh TT, Clemens MG: High-resolution visualization of oxygen distribution in the liver in vivo. *American Journal of Physiology-Gastrointestinal and Liver Physiology* 2004, 286:G37-G44.

142. Wang XF, Periasamy A, Herman B, Coleman D: Fluorescence lifetime imaging microscopy (FLIM): Instrumentation and applications. *Critical Reviews in Analytical Chemistry* 1992, 23:369-395.

143. Borisova EG, Troyanova PP, Avramov LA: Fluorescence spectroscopy for early detection and differentiation of cutaneous pigmented lesions. *Optoelectronics and Advanced Materials-Rapid Communications* 2007, 1:388-393.

144. Marcu L, Fang QY, Jo JA, Papaioannou T, Dorafshar A, Reil T, Qiao JH, Baker JD, Freischlag JA, Fishbein MC: In vivo detection of macrophages in a rabbit atherosclerotic model by time-resolved laser-induced fluorescence spectroscopy. *Atherosclerosis* 2005, 181:295-303.